# The Miracle of
# Baking Soda

Practical Tips for Health, Home, and Beauty

# The Miracle of
# Baking Soda

Dr. Penny Stanway

METRO BOOKS
New York

**METRO BOOKS**
New York

An Imprint of Sterling Publishing
387 Park Avenue South
New York, NY 10016

This 2013 edition published by Metro Books, by arrangement with
Watkins Publishing Limited

ISBN: 978-1-4351-4636-5

For information about custom editions, special sales, and
premium and corporate purchases, please contact Sterling Special
Sales at 800-805-5489 or specialsales@sterlingpublishing.com.

Designed and typeset by Tim Foster
Manufactured in China

2 4 6 8 10 9 7 5 3 1

www.sterlingpublishing.com

## *Acknowledgments*

Thank you to my sister, Jenny Hare, for her joy in food and cooking; to my husband, Andrew, for his unstinting enthusiasm in discussing baking soda and alkaline diets; to my family and friends for their good company and warm-heartedness when we eat together; to my agent, Doreen Montgomery, for her encouragement and support; and to my editor, Alison Bolus, for her patience, wisdom, and common sense.

### About the author

Dr. Penny Stanway practiced for several years as a doctor and as a child-health consultant before becoming increasingly fascinated in researching and writing about a healthy diet and other natural approaches to health and well-being. She is an accomplished cook who loves eating and very much enjoys being creative in the kitchen and sharing food with others. Penny has written more than 20 books on health, food, and the connections between the two. She lives with her husband in a houseboat on the Thames in London and often visits the south-west of Ireland. Her leisure pursuits include painting, swimming, and being with her family and friends.

# Contents

# Introduction

Baking soda is a white crystalline mineral powder with a multitude of uses, both domestic and commercial. It has the chemical code of $NaHCO_3$ and is also known as "bicarbonate of soda," "sodium bicarbonate," "sodium hydrogen carbonate," "bread soda," or simply "bicarb."

I have called it "baking soda" or "bicarbonate of soda" in those chapters dealing with cooking, home hints, and beauty, because this is the name most often used in these spheres of interest. However, in the rest of the book I have called it "sodium bicarbonate," because this is how it is known by physiologists, nutritionists, pharmacists, and doctors.

## KNOW YOUR SODA

The names of certain other chemicals could sound confusingly similar to those unfamiliar with chemistry. These include:

• baking powder (which contains other ingredients besides sodium bicarbonate)

• sodium chloride (salt)

• washing soda (sodium carbonate or soda ash)

• caustic soda (sodium hydroxide)

It is vital never to confuse any of these with sodium bicarbonate, so be sure that any container of white powder is labeled correctly and clearly.

Sodium bicarbonate is present (along with sodium carbonate) in the mineral known as natron, which is deposited from salt lakes in various countries. The ancient Egyptians used natron to clean their homes, their teeth and, when mixed with oil, their skin. They also used it to mummify dead bodies. Sodium bicarbonate can also be mined as deposits of nahcolite in rock, and it is present in the water flowing from many hot springs. Today, though, it is mostly produced in factories, mainly by a method that mixes salt (sodium chloride), ammonia, calcium carbonate, and carbon dioxide in water. A total of over 1 million tons is produced this way each year in various countries, including the USA, Italy, Egypt, Turkey, Russia, and China.

Sodium bicarbonate is mildly alkaline in water but is also amphoteric, which means it can react with both acids and alkalis. It is useful in many commercial and domestic situations. It is used, for example, as a leavening agent in cakes, breads, and batters, as well as to make baking powder, self-rising flour, soaps, domestic cleaning and deodorizing products, medications, water-softeners, "dry" fire extinguishers, and pesticides. It can also be used to make toothpaste and bath "bombs," and to coat dental floss.

Last, but not least, our body's own internally produced sodium bicarbonate is a vital part of the acid–alkali balancing systems that keep us healthy. And by eating an alkali-producing diet and, perhaps, taking sodium bicarbonate or applying it to our skin, we can help to treat certain common ailments.

This important mineral is available in small quantities as "baking soda" from supermarkets and grocery stores and as "sodium bicarbonate" from drugstores and pharmacies. You can also buy it in bulk online.

1

# Household
# Help

Baking soda is one of the most effective, inexpensive, and versatile non-toxic cleaning agents that you can use to keep your home and laundry fresh, clean, and fragrant.

# Homemade Cleaning Products

Long recognized as a highly effective cleaning agent, baking soda is mildly abrasive and has degreasing, water-softening, and deodorizing qualities. Keep it ready to use in a clearly labeled flour shaker. You can also use it to make the following cleaning products:

## Basic Cleaning Fluid

Put 2½ cups (21 fl oz, 600 ml) warm water into a large bowl and stir in 1 tablespoon of baking soda and two squeezes of dish detergent.

**Use and storage:** Make just enough for a particular cleaning job and use it from the bowl. Alternatively, pour it into a spray bottle, putting the top on only after any fizzing has stopped, and shake before each use.

## Extra-Strength Cleaning Fluid

To the **Basic Cleaning Fluid** recipe above, add either ½ cup (4 fl oz, 120 ml) white vinegar or the strained juice of two lemons. It will fizz a lot when first mixed.

**Use and storage:** Make just enough for a particular cleaning job and use it from the bowl. Alternatively, pour it into a labeled spray bottle, putting the top on only after any fizzing has stopped, and shake before each use.
Caution: This "Extra-Strength Cleaning Fluid" is unsuitable for waxed surfaces or granite or other stone surfaces, because vinegar and lemon juice can cause dulling.

# Basic Cleaning Paste

Put some baking soda into a bowl and stir in enough water to make a paste. The amounts required depend on the size of the job. The proportions are about three parts baking soda to one part water. Mix in a few drops of lemon, tea-tree, lavender, or other essential oil for fragrance, if wanted. This paste is excellent for cleaning areas of adherent grime, such as a ring around a bathtub.

**Use and storage:** Make just enough for a particular cleaning job and use it from the bowl. If the paste starts to dry, add more water.

# Extra-Strength Cleaning Paste

Make the **Basic Cleaning Paste** recipe above, but use vinegar instead of water.

**Use and storage:** Make just enough for a particular cleaning job and use it from the bowl. If the paste starts to dry, add more water.

# Cream Cleaner

Make the **Basic Cleaning Paste** recipe above, then stir in extra water until it has the consistency of thick cream. Add a few drops of lemon, tea-tree, lavender, or other essential oil for fragrance, if wanted.

**Use and storage:** Make enough for a particular job and use it from the bowl. Alternatively, pour it into a labeled squeezy bottle, putting the top on only after any fizzing has stopped, and shake before each use.

# Scouring Powder

This recipe includes borax substitute (sodium sesquicarbonate), which is available online (see page 181), and salt. The borax substitute has bleaching, stain-removing, deodorizing, and disinfecting qualities, while the salt has degreasing and cleaning qualities. Mix equal amounts of baking soda, borax substitute, and table salt in a bowl. Apply with a scrubbing brush, a cloth or, for small hard-to-reach areas, an old toothbrush.

**Use and storage**: Make enough for a particular job and use it from the bowl. Alternatively, put it into a labeled jar or plastic container.

## CLEANING PRECAUTIONS

If you are at all unsure about whether a cleaning product will be safe on a particular surface and what its effect might be, always try it out on a small, inconspicuous area first.

# How to clean ...

## Hands

Wash very soiled hands with soap, then sprinkle with baking soda. Rub your hands together, then rinse and dry.

Sprinkle baking soda into rubber gloves to keep them fresh and make them easy to put on.

## Sponges and cloths

Clean and deodorize sponges and cloths by soaking them in **Basic Cleaning Fluid** (see page 12), then rinsing and drying.

## Walls, doors, work surfaces, floors, and furniture

Baking soda helps to clean paint, laminate, other plastic, glass, granite, other stone, composite, rubber, brick, steel, fiberglass, and washable wallpaper. If using it on bare wood, be sparing with the water and dry well to avoid water marks.

- Mop or wipe floors with ½ cup (3½ oz, 100 g) baking soda dissolved in a bucket of hot water.

- Wipe other surfaces with **Basic Cleaning Fluid** (see page 12), then rinse.

- Rub particularly soiled or stained areas with baking soda sprinkled onto a damp sponge or cloth, then rinse.

- Banish grease or scuff marks on washable walls with baking soda sprinkled onto a damp sponge; rinse, then wipe dry.

# Dirty dishes

- Add ¼ cup (1¾ oz, 50 g) of baking soda and the juice of ½ a lemon to dishwashing water to aid degreasing and help loosen stuck-on food. You'll need less dish detergent than usual.

- Store steel-wool pan scourers in baking soda to prevent rust.

- Clean the kitchen sink with **Extra-Strength Cleaning Fluid** (see page 12) or rub all over with baking soda sprinkled onto a damp sponge or cloth.

# Sink pipes

- Help clear a block by pouring 1 cup (7 oz, 200 g) baking soda down the drain, followed by 1 cup (8 fl oz, 240 ml) hot vinegar. Wait for 30 minutes, then flush with a kettle of just-boiled water. If necessary, use a sink plunger.

- Freshen waste pipes and prevent them from becoming blocked by pouring ½ cup (3½ oz, 100 g) of baking soda down the drain and flushing with a kettle of just-boiled water each month.

# Dishwashers

- Mix 2 tablespoons of baking soda with 2 tablespoons of borax substitute (see page 181) to make dishwashing powder.

- Help prevent unwanted odors by filling the dishwasher-powder (or tablet) dispenser with baking soda and running a rinse cycle.

- Alternatively, clean inside and out with **Basic Cleaning Fluid** or **Extra-Strength Cleaning Fluid** (see page 12) and wipe clean.

- Or simply sprinkle ½ cup (3½ oz, 100 g) baking soda into the bottom of the dishwasher after emptying it.

## Refrigerators and freezers

- Clean a refrigerator or freezer by wiping inside and out with **Basic Cleaning Fluid** (see page 12), then rinsing. The remaining film keeps it smelling sweet.

- Rub any stained areas with **Basic Cleaning Paste** (see page 13), then rinse and dry.

- Make a refrigerator or freezer smell sweet by placing a bowl of baking soda inside. Stir it every few days and replace every 2–3 months.

## Wooden chopping boards

- Spring-clean a chopping board by sprinkling it with baking soda, then spraying it with white vinegar. Leave the bicarbonate-vinegar paste on for 30 minutes before rinsing with hot water.

- Deodorize a cutting board or wooden work surface that smells of garlic or onions by sprinkling some baking soda onto a damp sponge, rubbing it over the board, then rinsing it clean.

## Saucepans

Caution: Do not use baking soda on nonstick pans, as it can damage the nonstick surface.

Get rid of burnt-on food residue by:

- brushing or scrubbing a saucepan with baking soda or **Scouring Powder** (see page 14). Don't scour saucepans that have an enameled interior.

- wetting the saucepan with hot water then sprinkling on a thick layer of baking soda. Leave overnight, then brush or scrape off the residues.

- putting 1 cup (8 fl oz, 240 ml) water into the saucepan, adding 1 tablespoon white vinegar and bringing to the boil. Add 2 tablespoons baking soda and wait for the fizzing to subside, then brush or scrape off the residues.

## Oven and stove

Caution: **The "Basic Cleaning Paste"** (see page 13) can mark shiny stainless steel, and baking soda can darken aluminum or corrode the heating elements in an electric oven.

- Spray the oven walls with water, then spread them with a thick layer of **Basic Cleaning Paste** (see page 13). Leave this on for several hours, spraying every hour or so to keep the paste moist. Use a spatula, palette knife, or stove scraper to remove the debris. Wipe clean.

- Deal with charred greasy food residues on an oven floor by spraying with water, then sprinkling on a thick layer of baking soda and spraying again with water. Leave for several hours, spraying every hour or so to keep the paste moist. Wipe clean.

- If necessary, clean an oven or stove with **Extra-Strength Cleaning Paste** (see page 13).

- Clean a greasy stove, grill or splash-back by rubbing on baking soda with a damp sponge or cloth, then wiping clean.

- Degrease and clean an encrusted barbecue grill by applying **Extra-Strength Cleaning Paste** (see page 13), leaving for several hours, brushing with a wire brush, then wiping clean.

## Microwave oven

- Clean inside a microwave oven by wiping with **Basic Cleaning Fluid** (see page 12).

- Alternatively, put 1 tablespoon of baking soda and 1 cup (8 fl oz, 240 ml) water into a microwavable bowl and put this in the oven. Set the oven so the liquid boils for 3–4 minutes. Wipe the inside of the oven with a damp cloth or kitchen paper.

# Bathroom

- When you have a bath, add 2 tablespoons of baking soda to the water. This will help to prevent a scummy ring forming around the bathtub when you drain away the water.

- Clean the bathtub, basin, faucets, tiles, and mirrors by using a damp cloth or sponge to apply fragranced **Cream Cleaner** or **Basic Cleaning Paste** (see page 13). Rinse and dry.

- Alternatively, rub the surface with baking soda sprinkled onto a damp cloth or sponge, then rinse and dry.

- For heavier soiling or to clean glass shower screens, rub on **Extra-Strength Cleaning Paste** (see page 13), leave for 30 minutes, then rinse. Dry a glass shower screen with a barely damp towel.

- Help prevent mineral salts in hard water blocking a shower-head by pouring ¼ cup (1¾ oz, 50 g) baking soda and 1 cup (8 fl oz, 240 ml) white vinegar into a strong, hole-free plastic bag. Tie this around the shower-head and leave for 30 minutes. Remove, then run the water for a few seconds. (If you can remove the shower-head, simply immerse it in the above mixture for 30 minutes, then rinse.)

- Clean a toilet cistern and bowl by pouring 1 cup (7 oz, 200 g) baking soda into the cistern overnight and flushing next morning. Repeat once a month.

- Clean a toilet bowl by sprinkling baking soda onto a damp scrubbing brush and scrubbing under the rim.

- Treat a stained toilet bowl by rubbing with **Extra-Strength Cleaning Paste** (see page 13).

- Use an old firm-bristled toothbrush to apply **Extra-Strength Cleaning Paste** (see page 13) to soiled tile grouting. Leave to take effect for at least 30 minutes, then rinse.

- For particularly grimy or moldy grouting, mix three parts baking soda with one part bleach and scrub this over the grouting with an old toothbrush. Wait for 30 minutes, then rinse.

- Clean a soiled or mildewed machine-washable shower curtain by putting it in the washing machine along with a large bath towel. Add 1 cup (7 oz, 200 g) baking soda to the washing powder or liquid detergent. Then wash on a low-temperature setting, adding ½ cup (4 fl oz, 120 ml) of white vinegar to the fabric-softener dispenser during the rinse cycle. Don't spin fast, or the curtain could become permanently creased. Hang up to dry.

- Clean a non-machine-washable shower curtain by soaking it for 30 minutes in a bath of warm water containing about ¾ cup (5½ oz, 150 g) baking soda. Rinse and drip dry.

# Baby and child equipment and toys

- Dip children's dirty washable toys into **Basic Cleaning Fluid** (see page 12), or use a cloth to wipe them with the fluid. Rinse and dry.

- Wash a plastic paddling pool and remove any mildew with **Basic Cleaning Fluid** (see page 12).

- Clean high chairs, car seats, strollers, and plastic mattress protectors by sprinkling baking soda onto a damp sponge, rubbing with this, then wiping with a clean sponge several times.

- Remove milk residues from a baby's bottles, nipples, and bottle brushes by soaking them in **Basic Cleaning Fluid** (see page 12). Rinse, then sterilize in the usual way.

# Fruits and vegetables

Many people wash fruits and vegetables before eating them to get rid of dirt, pesticide traces, and micro-organisms. You could just use water, but a baking soda solution is better.

- Wash fruits and vegetables in a solution of 1 teaspoon of baking soda dissolved in a bowl of water, then rinse.

- Shake a little baking soda onto a wet vegetable brush and gently scrub firm fruits and vegetables.

- Clean soft fruits with a damp sponge sprinkled with baking soda.

# Metal

Baking soda can help to clean silver, chrome, and copper.

**Caution: Do not use baking soda on aluminum, as it would remove the thin protective coating of aluminum oxide. Bare aluminum reacts to acid so would soon look patchy if touched with sweaty hands or exposed to city air.**

- Scrub a silver item with an old toothbrush and some **Basic Cleaning Paste** (see page 12). Rinse with warm water, dry with kitchen paper, then polish with a soft cloth.

- Put tarnished silver on aluminum foil in a bowl of warm water containing 1 teaspoon of baking soda. (Or put the silver item into warm water in an aluminum container). Wait 5–15 minutes, then remove, rinse, dry with kitchen paper, and polish with a soft cloth. Silver reacts with oxygen and sulfur gases in the air to form a tarnish containing silver sulfide. In turn, aluminum reacts with silver sulfide to form aluminum sulfide, leaving sparkly bright pure silver.

- Clean chrome on car fenders and hubcaps, or on a bicycle, by rubbing baking soda over them with a damp sponge. Rinse, then polish with a soft dry cloth.

## DID YOU KNOW?

Baking soda was even used to clean the inner copper surface of the Statue of Liberty during restoration work for its centenniel celebrations in 1986.

- Make stainless steel shine by rubbing with a damp sponge sprinkled with baking soda, then rinsing and drying.

- Clean brass or copper objects by rubbing with baking soda sprinkled onto half a lemon. Rinse and dry.

- De-grime gold jewelry by putting it into a bowl, sprinkling it with baking soda, then pouring white vinegar over it. Rinse and dry. However, don't do this if the jewelry contains pearls or if gemstones have been glued in place rather than captured in metal "claws."

# Stain removing

**For stains on clothing, see "Laundering" (pages 26–28).**

- Get rid of tannin stains from tea and coffee by: using a damp sponge to rub baking soda onto cups and mugs, leaving for 30 minutes, then rinsing ● Filling cups and mugs with warm water and adding ½ teaspoon of baking soda, leaving for 30 minutes, then rinsing ● Filling a coffee pot or teapot with hot water and adding 2 teaspoons of baking soda and 2 teaspoons of white vinegar, leaving for 30 minutes, then rinsing ● Filling a glass and metal (but not aluminum) cafetière with hot water, stirring in 1 teaspoon of baking soda and 1 teaspoon of white vinegar, leaving for 30 minutes, then rinsing.

- Clean fruit drink or fruit juice stains from kitchen work surfaces or other washable surfaces by spraying them with water, then sprinkling them with baking soda. Leave for 30 minutes, then wipe clean.

- Remove crayon marks on a painted or papered wall by sprinkling baking soda onto a damp sponge, rubbing the wall very gently, then wiping clean.

- Eradicate water spots on wooden floors by dabbing with a sponge or cloth dampened with **Basic Cleaning Fluid** (see page 12). Wipe clean and repeat several times to remove all traces of baking soda. Dry well. Don't wet the wood too much, as this could simply make more water spots!

- Deal with a stained laminate (or other plastic), or a marble worktop or other surface by rubbing on some **Basic Cleaning Paste** (see page 13), then rinsing. Remember to try this out on a small, inconspicuous area if you are at all unsure of the results.

- Clean a stained vacuum flask or one that you haven't used for some time by putting 1–2 teaspoons of baking soda into the flask and filling it with hot water. Leave for 30 minutes, then rinse well.

- Remove wine, grease, and certain other awkward stains from carpets by lightly wetting the stains, then sprinkling them with baking soda. Leave to dry, then vacuum up the residue.

- Apply **Extra-Strength Cleaning Paste** (see page 13) to ink stains on hard floors, then rinse and dry.

- Clean brown staining on the base of your iron by unplugging and cooling it first, then rubbing with **Basic Cleaning Paste** (see page 13).

# Laundering

- Keep your laundry basket smelling fresh by sprinkling a little baking soda into it each day.

- Clean your washing machine drum, and inside the door, with **Basic Cleaning Fluid** (see page 12) or, for stubborn marks, **Basic Cleaning Paste** (see page 13).

- Deodorize your washing machine drum, and inside the door, by wiping with **Basic Cleaning Fluid** (see page 12).

- Alternatively, clean the washing machine by putting ¼ cup (1¾ oz, 50 g) of baking soda into the washing-powder dispenser and ¼ cup (8 fl oz, 240 ml) of white vinegar into the fabric-softener dispenser before running the machine on a short cycle.

- Add ½ cup (3½ oz, 100 g) of baking soda along with your usual amount of liquid washing detergent to the washing machine or hand-wash bowl for more effective washing power. This also helps to shift a variety of stains.

- Make a fabric conditioner by putting 1 cup (7 oz, 200 g) baking soda, 1 cup (8 fl oz, 240 ml) vinegar and 2 cups (16 fl oz, 480 ml) of water into a bottle large enough to accommodate the effervescence produced when baking soda mixes with vinegar. Add ¼ cup (2 fl oz, 60 ml) of this mixture to your washing machine's fabric-softener dispenser when you do a wash, or put it into the final rinse water if hand-washing. Baking soda helps to make the texture of towels soft and fluffy.

- For a fragranced fabric softener, use the above recipe but add 3–4 drops of lemon, lavender, or rose essential oil.

- Brighten up white clothes by adding ½ cup (3½ oz, 100 g) of baking soda to the washing-powder dispenser, or, if washing by hand, to the final rinse.

- Alternatively, add ¼ cup (1¾ oz, 50 g) of baking soda to a basin of cold water. Immerse the clothing and soak overnight, then wash as usual.

- For greater whitening action, add the juice of a lemon and ¼ cup (1¾ oz, 50 g) of baking soda to a basin of cold water. Soak the clothing overnight and wash it next morning.

- Boost the performance of bleach by adding ½ cup (3½ oz, 100 g) of baking soda and ½ cup (4 fl oz, 120 ml) of bleach to a bowl of water.

- Whiten cloth diapers by adding ½ cup (3½ oz, 100 g) of baking soda to the washing-powder dispenser, or along with the washing powder if you are washing by hand.

- Use baking soda to make chlorine bleach more effective: add ½ cup (3½ oz, 100 g) along with the usual amount of bleach.

- Deodorize soiled cloth diapers by putting ¼ cup (1¾ oz, 50 g) of baking soda into a bucket of cold water and soaking them overnight. Next morning, wash as usual.

- If clothing is stained with something acidic (such as fruit juice or tomato sauce), pre-treating it with baking soda before washing should prevent the acid eating into the fabric; it should also loosen the stain. Sprinkle with baking soda, spray with water then leave for 30 minutes before washing. This is also effective for removing acidic sweat, vomit, and urine stains.

- Remove a blood stain by putting ½ cup (3½ oz, 100 g) of baking soda and ½ cup (4 fl oz, 120 ml) of white vinegar into a bowl of cold water and soaking the item in the mixture overnight. Wash as normal next day.

- Get rid of grease on clothing by applying a paste of baking soda and water, leaving for 30 minutes, then adding ½ cup (3½ oz, 100 g) of baking soda to the washing machine (or a bowl if hand-washing) along with the liquid detergent. Run your usual cycle.

- Or make a paste from two parts of baking soda, one part of cream of tartar and a little water, and rub this on a grease mark before washing.

- Rub a tar stain with **Basic Cleaning Paste** (see page 13), then wash it with baking soda and water.

- Remove a rust stain from clothing by soaking it in lemon juice then sprinkling with a thick layer of baking soda. Leave overnight, then rinse and wash.

- Try eliminating a mildew smell from fabric by soaking it in **Basic Cleaning Fluid** (see page 12) overnight, then washing.

# Deodorizing

Baking soda neutralizes unpleasant smells. It acts against odors from acidic substances (such as in sweat, urine, or vomit) to alkaline ones (such as ammonia from wet diapers).

For refrigerators and freezers, see page 17. For dishwashers, see page 16. For wooden chopping boards, see page 17. For kitchen-sink pipes, see page 16. For laundry, see pages 26–28.

- Prevent unpleasant smells in a pantry or storecupboard by keeping a small open bowl of baking soda in it. Stir every few days and replace the baking soda every 2–3 months.

- Eliminate stale smells from plastic bowls or food containers by filling them with hot water and stirring in 1 tablespoon of baking soda. Soak for 30 minutes, then rinse.

- If the smell persists, repeat, adding 1 tablespoon of vinegar and a few drops of dish detergent along with the baking soda.

- Add a small handful of baking soda to keep a kitchen-garbage container smelling sweet.

- Prevent unpleasant smells from your garbage-disposal unit by pouring 2 tablespoons of baking soda and 1 tablespoon of white vinegar into it every week and running hot water from the faucet as you operate the disposer.

- Make a stored blanket smell sweeter by shaking baking soda over it, then rolling it up. Leave overnight, then shake it out next morning and run it in the tumble drier set to cold.

- Deodorize carpets and rugs by shaking baking soda all over them, waiting at least 30 minutes, then vacuuming.

- After cleaning spilt drink or food from a carpet, sprinkle the affected area with baking soda, wait 30 minutes, then vacuum. This makes any unpleasant smell less likely to linger.

- Deodorize the carpet in your car by sprinkling it with baking soda, leaving for at least 30 minutes, then vacuuming.

- Leave a small pot of baking soda in each room to act as an air freshener by neutralizing any unwanted odor. Commercial air fresheners mostly act by producing a masking scent.

- Make your own fragranced air freshener by adding a few drops of lemon, geranium, lavender, neroli, rose, or other essential oil of your choice to a small pot of baking soda.

- Neutralize a sour odor from cleaned-up human or pet vomit stains by sprinkling generously with baking soda. Leave for several hours, then vacuum the area.

- Sprinkle baking soda into smelly shoes, boots or trainers and leave overnight. Next morning, tap out the surplus. (Note that doing this with leather shoes could stiffen the leather.)

- Keep a closet smelling sweet by putting an open bowl of baking soda inside. Add a few drops of lemon, geranium, lavender, neroli, rose, or other essential oil for special fragrance.

- Alternatively, use scraps of fabric to make little bags which you can fill with baking soda and, perhaps, some fragrant essential oil. Tie up with string or ribbon and keep inside a closet or drawer.

- Sprinkle baking soda into garment storage bags to prevent musty smells.

- Freshen stuffed toys by sprinkling some baking soda over them. Wait for at least 30 minutes, then brush off.

- If there's a smoker in your home, put baking soda into the ashtrays to combat the stale tobacco smell.

- Sprinkle tents, waterproofs, and other camping gear with baking soda before storing.

- Rid a vacuum flask of stale smells by filling it with hot water and adding 2 teaspoons of baking soda. Soak for 30 minutes, then rinse.

- Help prevent odor from a pet's litter tray by putting a thick layer of baking soda on the bottom of the tray before adding the litter.

- In addition, sprinkle baking soda over a cat litter tray to help neutralize unpleasant smells.

- To remove the wet-dog smell from a damp dog's coat, sprinkle it with baking soda, wait 30 minutes, then brush it out.

- Make a dog or cat's bedding smell better by sprinkling it with baking soda, leaving it for 1 hour, then vacuuming.

## FIRE EXTINGUISHER

Baking soda can be used as a fire extinguisher as it releases carbon dioxide when heated, and so can smother flames.

If you don't have a fire extinguisher or a fire blanket at the ready, but you do have a large amount of baking soda, you can try putting out a small grease or oil fire, or an electrical fire, by throwing baking soda over it. (Water is unsafe for either sort of fire). Don't use baking soda on a fat-fryer fire, as it could make the flaming grease splatter.

Caution: take no risks and, if necessary, get yourself and others out of the house and make sure that someone calls the fire department.

# Rust removal

Help remove rust from metal objects by applying **Basic Cleaning Paste** (see page 13) with a damp cloth. Scrub lightly with a piece of aluminum foil, rinse, and dry with kitchen paper.

# Other

- Sprinkle baking soda onto floors or worktops or in cupboards to repel unwanted insects.

- Remove smears left after cleaning dead insects from a car windshield by applying baking soda on a damp sponge, then rinsing and wiping clean.

- Clean hairbrushes, combs, toothbrushes, and make-up applicators by soaking them in **Basic Cleaning Fluid** (see page 12) for 30 minutes, then rinsing well.

- Spray plant foliage affected by mildew fungi with **Basic Cleaning Fluid** (see page 12) in the evening.

- To treat plant foliage affected by black-spot fungi, add 1 teaspoon of cooking oil to a bottle of **Basic Cleaning Fluid** (see page 12), shake well to mix and spray in the evening.

- Clean vases inside by filling with hot water and adding 2 tablespoons of baking soda. Leave for 1 hour, then rinse.

# 2

# Cooking with Baking Soda

As its name suggests, baking soda is an essential kitchen storecupboard ingredient which has long been used as a leavening agent to add lightness and air to breads, cakes, batters, and other baked goodies. However, it has some other surprising culinary uses too.

Baking soda (also called bicarbonate of soda, bread soda, or cooking soda) is a godsend for any cook. One reason for this is that when combined with water and acid, it creates tiny carbon-dioxide bubbles that will aerate a cake mix, batter, or dough immediately, making it puffier and lighter. Once the mixture reaches a certain temperature in the oven, the chemical reaction ceases and the air bubbles become set in the mixture.

Yeast and eggs used to be the main leavening agents until the late 1700s, when scientists found that baking soda acted much faster.

Baking soda can be used:

- on its own, if the recipe contains an acidic food (see the list opposite) or cream of tartar (a powder that produces acid when mixed with water)

- in baking powder

- in self-rising flour

## ACIDIC FOODS THAT ENABLE LEAVENING
### BY BAKING SODA

Any of the following are commonly used in recipes with baking soda:

- Baking powder
- Beer
- Buttermilk
- Citrus juice
- Coffee
- Hard apple cider
- Honey
- Maple syrup
- Natural cocoa or chocolate (dark unsweetened varieties from cocoa beans that haven't been "Dutch-processed" to neutralize their acidity)
- Sour milk (bought from a shop, or made by stirring 1 tsp of vinegar or lemon juice into 1 cup of whole or 2 percent milk, and leaving for 15 minutes before using)
- Sour cream
- Molasses
- Vinegar
- Yogurt

# Cook's tips for using baking soda

## Do

- Store in an airtight container in a cool dry place.

- Sift with other dry ingredients several times before using to ensure thorough mixing.

- Mix wet and dry ingredients rapidly and cook a mixture without delay to prevent all the carbon-dioxide bubbles that were released during the mixing process from escaping.

- Use the amount recommended in a recipe. Too little can make a cake rise poorly and the finished product tough and dense. Too much can produce super-size bubbles that make a bread or cake rise too rapidly, which produces a coarse, open-textured loaf or cake and makes fruit and nuts sink. Also, the bursting bubbles make a cake sink in the middle.

- If cooking at high altitude, use less than a recipe suggests, as carbon-dioxide bubbles expand faster when air pressure is relatively low.

- Make chicken skin crispy by rubbing it with baking soda before cooking.

- If a sauce or casserole tastes too acidic, stir in ½–1 teaspoon of baking soda to reduce or neutralize the acidity.

- Tenderize pieces of beef, pork, or lamb by rubbing baking soda into the surface and leaving for a few hours (see page 54). Be sure to rinse off any residue and dry the meat on a paper towel before cooking.

- Soak dried beans in water containing baking soda. This makes them cook faster, reduces their content of starch and complex carbohydrates, which could ferment in the gut and produce gas, and makes them more digestible. Use 1 teaspoon of baking soda for each 1 cup (7 oz, 200 g) beans. Soak in a large saucepan of water for 12–24 hours, then rinse and cook.

- Baking soda loses its efficacy with time so do check that any old baking soda is still usable by mixing ¼ teaspoon with 2 teaspoons of vinegar—it should bubble at once.

**Don't**
- Add baking soda to green vegetables in order to retain their color when they are cooked because its alkalinity destroys vitamin C (ascorbic acid).

- Be tempted to use too much, as it could react with fat or oil to form traces of soap, giving the finished product a strange aftertaste!

- Substitute it with baking powder, as baking soda has four times the leavening power of baking powder.

# Baking powder

Baking powder contains baking soda, two acid-producing compounds, and cornstarch—which prolongs its shelf life by absorbing moisture, thus preventing the premature release of carbon dioxide.

One of the two acid-producing compounds in baking powder is chosen to be fast-acting. This means it produces acid as soon as it combines with water at room temperature in the mixing bowl. Examples include cream of tartar (potassium bitartrate) and calcium acid phosphate (mono-calcium phosphate).

The other acid-producing compound is chosen to be slow-acting. This means it produces acid only when heated in the oven or on the stove. Examples include sodium aluminum sulfate, sodium aluminum phosphate, and sodium acid pyrophosphate.

In this way, baking powder produces one batch of bubbles in the mixing bowl and another during cooking. It therefore has a double rising action.

So why don't cooks always use baking powder instead of baking soda? First, because many recipes contain an acidic food that will activate sodium bicarbonate. Second, because many recipes don't require a double rising action.

Interestingly, some recipes that call for baking powder still need extra baking soda to boost the rising action further, to reduce acidity, to lower the temperature at which sugar caramelizes, or to weaken gluten (a cereal-grain-flour protein) in order to produce a softer cake.

# Baking powder recipe

**To make baking powder with fast-acting rising ability: put 3 teaspoons of baking soda, 4 teaspoons of cream of tartar, and 1 teaspoon of cornstarch into a jar and shake well. Store in an airtight container.**

## Tips for using baking powder

- If short of an egg for a cake recipe, add an extra ½ teaspoon of baking powder plus 2 tablespoons of milk.

- Check that baking powder is usable by mixing 1 teaspoon with ½ cup (4 fl oz, 120 ml) hot water—it should bubble at once. Baking powder has a shelf life of 6–12 months and packs should display a use-by date.

## Self-rising flour

This is all-purpose flour combined with baking powder. However, it isn't always a suitable substitute for plain flour plus baking soda and an acidic ingredient, such as those listed on page 37. For example, when making a rich, dense cake, it is better to use all-purpose flour plus baking powder and, perhaps, extra baking soda. Some recipes made with self-rising flour also require extra baking soda.

# Self-rising flour recipe

**To make self-rising flour, for each 3¾ cups (1 lb, 450 g) all-purpose flour, add 2 teaspoons of baking soda and 4 teaspoons of cream of tartar. Sift all together three or four times and store in an airtight container.**

## PLEASE NOTE:

- Each recipe serves 4.

- 1 tsp (teaspoon) = 5 ml; 1 tbsp (tablespoon) = 15 ml.

- 1 cup = 8 fl oz or 240 ml.

- All fruit and vegetables are medium-sized unless otherwise stated.

- All eggs are large unless otherwise stated.

- If using a fan oven, reduce the temperature recommended in the recipe by 25°F/20°C.

- Salt is included only when needed to cure or soften other ingredients or to enhance their flavor. Anyone who wants to can add salt at the table.

# APPETIZERS, SNACKS, AND VEGETABLES

Baking soda softens dried beans and peas before they are used in such dishes as Fried Mung Beans and Dal. In combination with water and an acidic ingredient, it produces bubbles that puff up starchy ingredients such as wheat, ground cornmeal, polenta, or maize meal, or chickpea flour (also known as gram flour, garbanzo flour, and besan). This gives us such delights as Hush Puppies and Onion Bhajis.

# Fried Mung Beans

**If you cannot buy split dried mung beans, or mung dal, to make this snack, use split dried yellow peas instead. Amchoor is ground dried mango.**

*1 cup (8 oz, 225 g) split dried mung beans or yellow peas*

*1 tsp baking soda*

*olive oil, for frying*

*1 tbsp lemon juice*

*½ tsp ground cumin*

*½ tsp ground ginger*

*½ tsp coriander*

*½ tsp chili powder*

*ground black pepper*

*½ tsp amchoor (optional)*

- Put the mung beans into a large bowl of water, add the baking soda and soak for 8 hours, topping off with more water if necessary.

- Heat plenty of oil in a frying pan or skillet. Fry the drained beans until whitish and crisp. Remove and drain on kitchen paper. Put them into a bowl and stir in the lemon juice, spices, pepper, and amchoor, if using. Serve hot or cold.

# Hush Puppies

**These savory Creole corn-batter puffs make a good starter and are great served with fish or meat.**

*oil, for deep-frying*
*1 cup (5 oz, 140 g) ground cornmeal or polenta*
*heaped ¾ cup (4 oz, 100 g) all-purpose flour*
*2 tsp baking powder*
*1 tsp cayenne pepper*
*ground black pepper*
*3 scallions, chopped*
*¼ onion, finely chopped*
*½ green bell pepper, finely chopped*
*1 egg, beaten*
*1 cup (8 fl oz, 240 ml) milk*

- Preheat the oil to 375ºF (190ºC). Preheat the oven to 300ºF (150ºC, Gas Mark 2).

- Put the cornmeal, flour, baking powder, cayenne pepper, and a good grinding of black pepper into a large bowl and mix well. Stir in the scallions, onion, and bell pepper. Add the beaten egg and half the milk and stir well, then add enough extra milk to form a batter that is stiff but that you can drop from a spoon.

- Drop several tablespoons of batter separately into very hot oil and fry for 2–3 minutes or until light golden brown all over, turning them several times as they puff up. Drain on kitchen paper and keep warm in the oven. Cook the remaining batter in batches in the same way.

# Onion Bhajis

**Originating in India, these spicy fried balls of onion, spinach, and chickpea (gram) flour are popular around the world.**

*2 tbsp all-purpose flour*
*1½ tsp baking soda*
*1 tsp chili powder (or paprika for milder bhajis)*
*½ tsp ground turmeric*
*scant 2 cups (7 oz, 200 g) chickpea flour*
*2 cups (2 oz, 50 g) spinach, chopped*
*1 onion, finely chopped*
*oil, for deep-frying*
*mango chutney or sweet chili dipping sauce, to serve*

- Sift the all-purpose flour, baking soda, chili, and turmeric into a large bowl. Stir in the chickpea flour, spinach, and onion. Add enough water to bind the mixture. Leave for 30 minutes.

- Heat the oil in a pan until a breadcrumb dropped into it sizzles and browns. Drop in tablespoons of the chickpea mixture and cook, turning occasionally, for 3–4 minutes or until golden-brown.

- Remove the bhajis with a slotted spoon and drain on kitchen paper. Serve hot with mango chutney or sweet chili dipping sauce.

# Mushy Peas

**A comforting dish for a cold winter's day, this is served as a traditional accompaniment to fish and chips and, in the north of England, to baked ham.**

*2 tsp baking soda*
*heaped 1 cup (12 oz, 350 g) dried marrowfat peas*
*¼ stick (1 oz, 30 g) butter*
*ground black pepper, to taste*

- Soak the peas in a large bowl of water with the baking soda for 4–8 hours, or overnight. Drain the peas, rinse under running water, put into a large pan and cover with fresh water. Cover with a lid and bring to the boil, then simmer for 1½–2 hours, stirring occasionally.

- Drain, then stir in the butter and ground pepper.

# Dal

**Also spelt "daal" and "dahl," this spicy bean and lentil purée can be scooped up with naan bread (see page 68).**

*1 cup (8 oz, 225 g) dried red kidney beans, soaked overnight in water with 1 tsp baking soda*

*1 cup (6 oz, 175 g) dried black lentils*

*3 tbsp oil*

*2 onions, finely chopped*

*2 tomatoes, finely chopped*

*2 garlic cloves, crushed*

*1 in (2.5 cm) root ginger, grated*

*1 tsp garam masala*

*1 tsp chili powder*

*1 tsp turmeric*

*2 tbsp cream*

*¼ stick (1 oz, 30 g) butter*

*1 large handful cilantro, finely chopped*

- Drain and rinse the kidney beans, then put into a large pan of water, boil for 1¼ hours, then drain. Meanwhile, put the lentils into another pan of water, boil for 30 minutes, then drain.

- Heat the oil in a large frying pan or skillet. Add the onions and fry gently until lightly browned. Stir in the tomatoes, garlic, and ginger and fry gently for 5 minutes. Stir in the garam masala, chili powder, turmeric, beans, and lentils and fry gently for a further 5 minutes. Stir in the cream, then remove from the heat. Put the butter on top of the dal so it melts, then garnish with cilantro.

# MAIN COURSES

Baking soda is an important ingredient for softening dry beans and pulses, especially when preparing falafels or similar dishes that require pounding the beans down. It also adds airy lightness to savory batters and acts as an effective meat-tenderizer, making it particularly useful when preparing beef, pork, or lamb for stir-frying.

# Falafels

**These herb-flavored delicacies originated in Egypt but are eaten throughout North Africa, the Middle East, and in many other countries too. They can also be made from chickpeas.**

*2 cups (11 oz, 300 g) dried fava beans*
*2 tsp baking soda*
*small handful of dill*
*small handful of cilantro, chopped*
*small handful of parsley, chopped*
*½ green bell pepper, seeded and chopped*
*2 onions, finely chopped*
*4 garlic cloves, crushed*
*1 tsp ground cumin*
*ground black pepper*
*½ tsp cayenne pepper*
*1 egg*
*2 tbsp sesame seeds*
*oil, for deep-frying*
*pita breads and salad, to serve*

## For the yogurt dip

*1⅔ cups (14 oz, 400 g) natural yogurt*
*6 in (15 cm) piece of cucumber, chopped*
*½ onion, chopped*
*1 garlic clove, crushed*
*1 tbsp mint leaves, chopped*

- Rehydrate the beans by soaking them overnight in a large pan of water containing the baking soda. Next day, drain the beans, put them into a large bowl and mash them down or pulse in a blender.

- Add the dill, cilantro, parsley, green bell pepper, onions, garlic, cumin, black pepper, cayenne, and egg and mix to form a paste. Knead for 2 minutes. Cover with a dishtowel and leave for 30 minutes.

- With floured hands, shape a little of the mixture into a ball 1½ in (4 cm) across. Make balls from the remaining mixture. Put the sesame seeds onto a plate and roll each ball in the seeds to coat it.

- Heat the oil in a large pan until a cube of bread dropped in sizzles and browns. Deep-fry several balls at a time for 3–4 minutes or until golden brown, turning several times.

- Remove the falafels with a slotted spoon and drain on kitchen paper. Repeat with the remaining balls.

- Make the yogurt dip by mixing all the ingredients together in a bowl.

- Serve the falafels, hot or cold, with the yogurt dip, along with pita breads and a green salad.

# Potato Gnocchi

**This popular Italian dish makes a good main course if sprinkled with parmesan cheese and served with salad. Alternatively, serve it as a side with meat or fish.**

*1½ lb (675 g) potatoes, peeled and cut into chunks*
*1 cup (5 oz, 150 g) all-purpose flour, plus extra for dusting*
*1 tsp baking powder*
*1 egg, beaten*
*1 tbsp olive oil*
*pinch of salt*
*good grinding of black pepper*
*2 tsp of fresh sage, chopped, or 1 tsp of dried sage*

- Put the potatoes into a large pan of cold water, bring to the boil, and simmer for 15 minutes or until just tender. Drain, then mash. Sieve the flour and baking powder over the mashed potatoes, then stir in. Add the egg and olive oil and mix to make a smooth dough.

- Knead the dough on a lightly floured board, dusting with extra flour if necessary to prevent sticking. Roll a small piece of dough into a ball the size of a cherry tomato. Flatten and "groove" it with the back of a fork, then fold it in half and set aside. Repeat with the remaining dough.

- Bring a large pan of salted water to the boil. Add the gnocchi. When they float toward the top, cook for a further 2 minutes. Drain, then sprinkle with the sage.

# Tempura

**Tempura batter makes a thin, light, and crispy coating for chicken, fish, and vegetables. Allow some small lumps of flour to remain, as these give the cooked batter its traditional texture.**

*⅔ cup (3 oz, 75 g) all-purpose flour, plus extra for dusting*
*1 cup (5 oz, 125 g) cornstarch*
*2 tsp baking powder*
*ground black pepper*
*generous 1 cup (9 fl oz, 270 ml) ice-cold sparkling water*
*1 lb 8 oz (675 g) chicken breast or fish fillet, cut into ½ in (1 cm) pieces, or 2 eggplants,*
*or 4 zucchini, cut into ½ in (1 cm) pieces, or 1 lb (450 g) small mushrooms*
*olive oil, for deep-frying*

- Put the flour, cornstarch, baking powder, black pepper, and water into a large bowl and mix together lightly.

- Heat the olive oil to 375°F (190°C). Dust a few pieces of chicken, fish or vegetable with flour, then lower them into the batter and deep-fry for 2–3 minutes or until light golden brown and crispy. Serve at once.

# Oh-So-Tender Beef Stir-Fry

**Rubbing baking soda into sliced meat makes it wonderfully tender for a stir-fry.
If you prefer, substitute pork or lamb for the beef.**

*1½ lb (680 g) beef, cut into thin strips*
*2 tsp baking soda*
*2 tbsp roasted sesame, or peanut oil*
*2 onions, finely sliced*
*good grinding of black pepper*
*3 garlic cloves, crushed*
*1 large red chili, finely chopped*
*2 in (5 cm) piece of fresh ginger root, grated*
*2 cups (16 fl oz) bean sprouts*
*2 tsp soy sauce*
*small handful of cilantro or parsley leaves*
*1 lime, quartered*

- Lay the beef in a dish. Mix the baking soda with 2 tablespoons of water and rub it into the beef. Cover the dish with plastic wrap, and leave in the refrigerator for 1–2 hours. Then rinse the beef well and dry thoroughly with kitchen paper.

- Heat the oil in a large frying pan or a wok. Add the onions and fry, stirring occasionally, until they begin to soften. Turn up the heat and add the pepper, beef, garlic, chili, and ginger. Continue frying, stirring frequently, for 2–3 minutes, until the beef is cooked. Now add the bean sprouts and fry, stirring frequently, for 2–3 minutes or until they wilt. Sprinkle with the soy sauce, cilantro or parsley, and serve with the lime on the side.

# DESSERTS, CAKES, AND COOKIES

Baking soda and baking powder come into their own when making light and airy treats such as apple fritters and sweet cookies. And almost every cake recipe contains baking soda either as the sole leavening agent or as a component of baking powder.

# Apple Fritters

**Slices of peeled and cored apples coated in a light batter and deep-fried in hot oil make wonderful fritters.**

*heaped ¾ cup (4 oz, 100 g) all-purpose flour*

*1 tsp ground cinnamon*

*1 tbsp olive oil*

*1 egg, separated*

*about ⅓ cup (2½ fl oz, 70 ml) milk*

*1 tsp baking powder*

*4 large sweet dessert apples, peeled, cored, and cut into ¼ in (6 mm) rings or slices*

*oil, for deep-frying*

*sugar, to sprinkle*

*vanilla ice cream, to serve*

- Sift the flour and cinnamon into a bowl. Stir in the olive oil, egg yolk, and enough milk to make a batter that just coats the back of a wooden spoon. Leave for half an hour. Immediately before using it, whisk the egg white. Then stir the baking powder into the batter and fold in the whisked egg white.

- Heat the oil to 375ºF (190ºC). Dip a few apple rings or slices into the batter then drop them into the hot oil. Turn them several times and, when golden-brown, remove with a slotted spoon and drain on kitchen paper. Repeat with the remaining slices of apple.

- Sprinkle with sugar and serve warm with vanilla ice cream.

## VARIATIONS
- For a spicy batter, add 1 teaspoon of ground cinnamon to the flour.
- Instead of apple slices, try fritters made with slices of fresh or canned pineapple, or bananas cut into 2 in (5 cm) chunks in place of apple.

# Semolina Cakes

These cakes, made with semolina and soaked in syrup or honey, have a very long history in India, Pakistan, Bangladesh, Afghanistan, Iran, and Greece, Turkey and other Mediterranean countries. They are also known as "semolina halva," but their distinctive and delightful texture is less dense than that of traditional halva made with tahini (sesame paste).

*1 stick (4 oz, 115 g) butter*
*¼ cup (2 oz, 50 g) sugar*
*1 tsp vanilla extract*
*2 eggs, beaten*
*¾ cup (6 oz, 175 g) plain yogurt*
*3⅓ cups (14 oz, 400 g) semolina*
*1 tsp baking powder*
*½ tsp baking soda*
*12 blanched split almonds*

### For the syrup
*scant 2 cups (14 oz, 400 g) sugar*
*1 tbsp lemon juice*

- Preheat the oven to 350°F (180°C, Gas Mark 4). Grease an 8 x 12 in (20 x 30 cm) shallow baking pan.

- Put the butter, sugar, and vanilla extract into a large bowl and beat well until light and creamy in color. Gradually beat in the eggs one at a time, then add a little of the yogurt.

- Sift the semolina, baking powder, and baking soda twice into another large bowl. Fold this dry mixture into the egg mixture a little at a time, alternating with the remaining yogurt. Pour the combined mixture into the baking pan and decorate with rows of almonds on top. Bake for 30–35 minutes or until a skewer inserted into the cake comes out clean.

- To make the syrup, put the sugar, lemon juice, and scant 1½ cups (12 fl oz, 350 ml) water into a small saucepan. Bring to the boil, stirring constantly, and boil rapidly for 6 minutes, ensuring that the mixture doesn't burn or stick. Remove from the heat and cool the pan by standing it in cold water.

- Once the cake is cooked, spoon the cooled syrup over it. Once cold, cut it into diamonds or squares.

# Devil's Food Cake

Baking soda reddens natural cocoa (which is slightly acidic) during baking, hence the name "devil's food cake." Natural cocoa powder is neither Dutch-processed (treated with an alkali to neutralize the acids) nor sweetened. You can also produce a fine cake with Dutched cocoa powder as long as the recipe contains an acidic ingredient such as coffee (as below). Note that it doesn't matter which type of milk you use.

*1 stick (4 oz, 100 g) unsalted butter, at room temperature, plus extra for greasing*
*1¼ cups (5 oz, 150 g) all-purpose flour*
*scant ⅔ cup (2½ oz, 75 g) cocoa powder, preferably natural*
*1 tsp baking soda*
*¼ tsp baking powder*
*½ cup (4 fl oz, 120 ml) strong coffee*
*½ cup (4 fl oz, 120 ml) milk*
*1½ cups (12 oz, 340 g) sugar*
*2 extra large eggs, at room temperature*

## For the frosting:

*10 squares (10 oz, 275 g) semisweet chocolate, coarsely chopped*
*1½ sticks (5 oz, 150 g) unsalted butter, finely chopped*

- Preheat the oven to 350ºF (180ºC, Gas Mark 4). Grease and line two 9 in (23 cm) round cake pans.

- Sift the flour, cocoa powder, baking soda, and baking powder into a large bowl. Pour the coffee and milk into a small bowl and stir well.

- Put the butter and sugar into another large bowl and beat with a wooden spoon for at least 5 minutes, until light and creamy. Alternatively, beat with a standing electric mixer. Add one of the eggs and a tablespoon of the flour mixture and beat well. Repeat with the other egg.

- Stir in half the flour mixture, then the coffee and milk mixture, then the remaining flour mixture. Pour the batter into the prepared pans and bake for 25 minutes or until a skewer inserted into the middle comes out clean. Leave to cool for a few minutes, then turn out onto a wire rack and leave to cool completely.

- To make the frosting, melt the chocolate and ½ cup (4 fl oz, 120 ml) water in a bowl over a pan of simmering water, stirring occasionally. Remove from the heat, add the butter and stir until melted. Leave to cool for 1 hour or until the frosting is thick enough to be spread.

- Place one cake on a plate and spread a third of the frosting over the top of it. Put the other cake on top of this and spread the top and sides of the whole cake with the remaining frosting.

# Apple Cider Cake

**The hard cider in this recipe is just acidic enough to activate the baking soda. This produces bubbles of carbon dioxide, which make the cake rise.**

*scant 2 cups (8 oz, 225 g) all-purpose flour*
*1 tsp baking soda*
*1 tsp ground nutmeg*
*1 stick butter (4 oz, 100 g) softened, plus extra for greasing*
*scant ½ cup (4 oz, 100 g) sugar*
*2 eggs, beaten*
*generous ½ cup (5 fl oz, 150 ml) sweet hard cider*

- Preheat the oven to 350°F (180°, Gas Mark 4). Grease and line a 7 in (18 cm) square baking pan.

- Sift the flour, baking soda, and nutmeg into a bowl.

- Put the butter and sugar into a large bowl and beat with a wooden spoon until smooth and pale. Add 1 tablespoon of the beaten eggs and 1 tablespoon of the flour mixture and beat well. Repeat, then beat in the remaining egg.

- Add the remaining flour mixture and stir well. Then add the hard cider and beat the mixture until it begins to froth.

- Pour into the baking pan and bake for 40 minutes or until the cake shrinks from the sides of the tin. Turn out onto a wire rack to cool.

# Festive Gingerbreads

**In Scandinavia, these spiced cookies are made in the run-up to Christmas.**

*⅔ cup (8 oz, 225 g) corn syrup*
*scant ¾ cup (5 oz, 150 g) sugar*
*1 tsp ground ginger*
*1 tsp ground cinnamon*
*1 tsp ground cloves*
*1¼ sticks (5 oz, 150 g) butter*
*1½ tsp baking soda*
*1 egg, beaten*
*4¼ cups (1 lb 4 oz, 550 g) all-purpose flour, plus extra for dusting*

- Put the syrup, sugar, spices, and butter into a large saucepan and heat gently, stirring, until the sugar and butter have melted. Leave to cool for 15 minutes. Dissolve the baking soda in 2 teaspoons of water and add to the pan. Add the egg, sift in the flour and mix together.

- Transfer the resulting dough to a bowl and leave in a cool place for 1 hour to make it easier to roll out.

- Preheat the oven to 425°F (220°C, Gas Mark 7). Grease four cookie sheets. Put the dough on a floured board and roll to about ⅛in (3 mm) thick. Cut out gingerbread-men, stars, or other shapes and place on the cookie sheets.

- Bake for 10–15 minutes until golden. Then transfer onto wire racks to cool.

# Anzac Cookies

**These "soldiers' biscuits" eased the long voyage endured by the Australian and New Zealand Army Corps (ANZAC) who fought in Turkey in the First World War.**

*½ cup (2 oz, 50 g) all-purpose flour*
*½ cup (2 oz, 50 g) wholemeal flour*
*1 cup (3 oz, 80 g) rolled oats*
*½ cup (1½ oz, 40 g) desiccated coconut*
*1 cup (7 oz, 200 g) brown sugar*
*1 ¼ stick (5 oz, 125 g) butter*
*2 tbsp dark corn syrup*
*1 tbsp boiling water*
*½ tsp baking soda*

- Preheat the oven to 350°F (180°C, Gas Mark 4). Grease two 16 x 12 in (40 x 30 cm) cookie sheets.

- Sift the plain and wholemeal flours into a large bowl. Stir in the rolled oats, desiccated coconut, and brown sugar.

- Put the butter into a saucepan and heat gently, stirring, until melted. Add the corn syrup and boiling water, and stir until the syrup has dissolved. Remove from the heat and stir in the baking soda. Stir this mixture into the dry ingredients.

- Divide the resulting dough into about 24 walnut-sized pieces, then roll these into balls. Place on the cookie sheets and flatten them. Bake in the center of the oven for 15–20 minutes until golden. Remove and leave to cool on the sheets.

# BREADS AND CRACKERS

Baking soda makes a rapid-acting alternative to yeast as a leavening agent for a wide variety of breads and crackers.

# Naan Bread

**This flat bread is a staple in South and Central Asia and is wonderful for mopping up sauces. Add one or more of the optional flavorings to complement your meal.**

*heaped 2 cups (9 oz, 250 g) all-purpose wholewheat flour, plus extra for dusting*

*2 tsp sugar*

*½ tsp baking powder*

*½ tsp baking soda*

*½ cup (4 fl oz, 120 ml) milk*

*2 tbsp olive oil, plus extra for greasing*

*3 tbsp yogurt*

*2 garlic cloves, chopped and lightly fried in oil or butter (optional)*

*1 tbsp cilantro, chopped (optional)*

*⅓ cup (2 oz, 50 g) poppy or sesame seeds (optional)*

*1 tbsp (½ oz, 15 g) butter, melted*

- Sift the flour, sugar, baking powder, and baking soda into a large bowl and make a well in the center. Pour the milk, oil, and yogurt into another bowl and stir well, then pour this mixture into the well. Add the chopped garlic or cilantro, if using.

- Turn the mixture out onto a floured surface and knead for 10 minutes, adding a little more flour if the dough is too sticky.

- Put the dough into a greased bowl, cover with a damp cloth and leave in a warm place for 2–3 hours. Shape into five balls.

- Preheat the oven to 425°F (220°C, Gas Mark7). Put two cookie sheets into the oven to heat up.

- Roll out the dough balls thinly into teardrop shapes. Sprinkle poppy or sesame seeds onto the flattened dough shapes, if using, and press into their surface.

- Put the breads on the hot cookie sheets and bake for about 10 minutes or until slightly puffed up and light golden brown.

- Pour the melted butter over the naan breads and serve hot.

## VARIATION
## PESHWARI NAANS

*4 tsp sugar*
*2 tbsp ground almonds (optional)*
*1 tbsp desiccated coconut, soaked in water then squeezed out (optional)*
*2 tbsp golden raisins*

- For sweet Peshwari naans, make the naan dough as in the main recipe.

- Mix the sugar, ground almonds (if using), coconut (if using), and golden raisins together. Slightly flatten each dough ball, put some filling into the center, then bring the sides over the mixture to stick together in the middle. Roll into shapes gently so that the filling does not spill out. Bake in the oven as above.

# Cornbread

Traditional in the American south, this bread gains its light and gritty texture from the cornmeal (also known as maize meal and polenta) and looks like a tray-bake. When combined with chopped onions and sage gently fried in butter, it also makes an excellent stuffing for a turkey.

*1 stick (4 oz, 100 g) butter, melted, plus extra for greasing*

*1½ cups (6 oz, 175 g) all-purpose flour*

*1 tbsp baking powder*

*scant 2 cups (10 oz, 275 g) cornmeal*

*2 tbsp sugar*

*4 eggs, beaten*

*1 cup (8 fl oz, 240 ml) milk*

*scant ¼ cup (2 fl oz, 50 ml) half and half cream*

- Preheat the oven to 400ºF (200ºC, Gas Mark 6). Grease an 8 in (20 cm) square heavy baking dish and put it in the oven.

- Sift the flour and baking powder into a large bowl. Stir in the cornmeal and sugar. Pour in the butter and stir until the mixture resembles breadcrumbs.

- Put the eggs and milk into another bowl and whisk. Stir into the flour and butter mixture. Then stir in the cream to make a paste-like mixture.

- Take the warmed baking dish from the oven and spoon in the mixture. Bake for 25–30 minutes or until a skewer inserted into the center comes out clean. Leave to cool for 5 minutes. Serve hot or cold in large squares or slices.

# Irish Soda Bread

**This was the first recipe I learned at school from our cooking teacher Mrs Stallebrass. She said it was quicker to cook than yeast-leavened bread, so if we could make it we'd never be short of bread for unexpected guests. Soda bread originated in Ireland, and in south-west Ireland it is often called "cake." Eat with butter and jam, or soup, or use it to mop up gravy or another sauce.**

*2¾ cups (14 oz, 400 g) wholewheat flour*
*heaped ¾ cup (4 oz, 100 g) all-purpose flour*
*scant ½ cup (2½ oz, 60 g) cornmeal*
*1 tsp baking soda*
*2–3 cups (16–24 fl oz, 480–720 ml) buttermilk (or milk plus an extra ½ tsp baking powder)*

- Preheat the oven to 450ºF (230ºC, Gas Mark 8). Dust a cookie sheet with flour. Sift the flours, cornmeal, and baking soda into a large bowl. Make a well in the center and quickly stir in three-quarters of the buttermilk, adding more if necessary to make a very soft, even slightly sticky dough. Put this onto a floured surface and knead quickly (any longer would allow carbon-dioxide bubbles to escape and thereby toughen the bread).

- Put the dough onto the cookie sheet and shape into a mound 6–8 in (15–20 cm) in diameter. Mark a cross over the top with a knife; the cuts should go halfway down the sides so the loaf will "flower" as the dough expands in the oven.

- Bake for 10 minutes, then reduce the heat to 400ºF (200ºC, Gas Mark 6) and bake for another 35 minutes. For a crunchy crust, cool on a rack uncovered; for a softer crust, wrap the loaf in a clean, damp cloth after removing it from the oven.

# Oatcakes

These savory crackers are delicious with cheese, peanut butter, or jam. Note that oatmeal is made by grinding whole dehusked oats to a fine, medium, or coarse grade of meal. Steelcut oats are made by cutting rolled oats, and rolled oats are made by rolling whole oats, but neither type has a suitable texture for this recipe. Look for a coarse "Scottish" oatmeal (available online, see page 180), which will produce a coarse-textured oatcake. However, whizzing it in a food processor for a short time will result in texture approximate to that of "medium" oatmeal.

*scant ⅔ cup (3 oz, 75 g) self-rising flour*
*scant ½ cup (4 oz, 100 g) medium oatmeal*
*½ tsp baking powder*
*heaped ¾ cup (2 oz, 4 tbsp, 50 g) white vegetable shortening*
*flour, for dusting*

- Preheat the oven to 400ºF (200ºC, Gas Mark 6). Grease a cookie sheet.

- Put the flour, oatmeal, and baking powder into a large bowl and mix well. Lightly cut in the vegetable shortening until the mixture is crumbly. Gradually add enough cold water to form a stiff dough.

- Put the dough onto a lightly floured surface and roll it out to ¼ in (5 mm) thick. Use a large round plain cutter (or a teacup) to form 8–10 circles. Put these onto the cookie sheet and bake in the oven for about 20 minutes, taking care that the oatcakes do not brown. Transfer to a wire rack to cool.

# SCONES, PANCAKES, AND WAFFLES

Batters aerated with a leavening agent give us some of lightest and most popular and adaptable dishes, including waffles, pancakes, and scones, which are all good served with sweet or savory accompaniments.

# Scones

**Warm scones with butter and jam are a treat fit for the gods. Some people also enjoy them with cheese and chutney. If you want to make buttermilk scones, omit the cream of tartar.**

*¼ stick (1 oz, 30 g) chilled butter, diced, plus extra for greasing*
*scant 2 cups (8 oz, 225 g) all-purpose flour, plus extra for dusting*
*1 tsp baking soda*
*2 tsp cream of tartar (optional)*
*scant ½ cup (2 oz, 50 g) golden raisins (optional)*
*generous ½ cup (5 fl oz, about 150 ml) milk or buttermilk*

- Preheat the oven to 230ºC (450ºF, Gas Mark 8). Lightly grease a cookie sheet.

- Sift the flour, baking soda, and cream of tartar (if using) into a large bowl and lightly cut in the butter. Stir in the raisins (if using). Immediately mix in enough of the milk or buttermilk to make a soft dough, using a round-bladed knife.

- Turn the dough out onto a lightly floured surface and knead quickly until smooth. Roll out the kneaded dough to ¾ in (2 cm) thick and cut out 8–10 rounds with a 2 in (5 cm) cutter. Place on a cookie sheet and brush the tops with milk.

- Bake the scones for about 10 minutes, until well risen and golden brown. Transfer to a wire rack to cool.

# Singing Hinnies

**These fried scones are teatime treats that originated in the north of England. Affectionately known as "hinnies" ("hinny" is local slang for "honey"), they delight the ear by whistling or "singing" as they cook.**

*2 cups (8 oz, 225 g) scant all-purpose flour*
*½ stick (2 oz, 50 g) butter, plus extra for frying*
*4 tbsp (2 oz, 50 g) white vegetable shortening*
*¼ cup (1 oz, 30 g) chopped raisins*
*1 tsp baking powder*
*about 2 tbsp milk*
*butter and jam or corn syrup, to serve*

- Put the flour, butter, and shortening into a large bowl and cut in the fats until the mixture resembles fine breadcrumbs. Add the raisins, baking powder, and milk and stir to form a soft dough.

- Roll the dough out on a floured surface to form a circle about ¾ in (2 cm) thick. Cut into 6–8 wedges.

- Melt some butter in a large frying pan or skillet, reduce the heat and cook the circle of wedges on both sides until lightly browned.

- Split and serve with butter and jam or corn syrup.

# Scotch Pancakes

**These small pancakes are made from a thick pancake batter puffed up a little with baking soda. Their other name is "drop scones." Buttermilk is a good alternative to the milk, but as it is more acidic you should then halve the amount of cream of tartar.**

*scant 2 cups (8 oz, 225 g) all-purpose flour*
*1 tsp baking soda*
*1 tsp cream of tartar*
*⅓ cup (3 oz, 75 g) sugar*
*2 eggs*
*1¼ cups (10½ fl oz, 300 ml) milk*
*butter, for frying*

- Sift the flour, baking soda, and cream of tartar into a large bowl. Stir in the sugar. Make a well in the center and whisk in the eggs and enough milk to make a smooth batter.

- Melt a knob of butter in a frying pan or skillet, or on a griddle. Spoon tablespoonfuls of batter into the pan, leaving enough room for them to spread. When the pancakes puff up and bubbles start bursting on their surface, turn them over using a palette knife. Cook until golden-brown underneath.

- Remove the pancakes from the heat, place them on a dishtowel and cover with another dishtowel to keep them warm and prevent them drying out while you cook the next batch.

# Waffles

**To make these vanilla-flavored waffles, you will need either an electric waffle iron or one that you heat on the stove. If you wish to make unsweetened waffles to eat with bacon and scrambled eggs or grated cheese, omit the sugar and vanilla.**

*heaped ¾ cup (4 oz, 100 g) all-purpose flour*
*2 tsp baking powder*
*2 tbsp sugar*
*1 egg, separated*
*¼ stick (1 oz, 30 g) unsalted butter, melted, plus extra for greasing*
*½ tsp vanilla extract*
*about 1 cup (8 fl oz, 240 ml) milk*
*maple syrup, corn syrup, jam, butter, or ice cream, to serve*

- Sift the flour and baking powder into a large bowl and stir in the sugar. Make a well in the center and drop in the egg yolk, melted butter, and vanilla extract. Gradually beat in enough of the milk to make a smooth batter with the consistency of thin cream. Whisk the egg white until stiff, then fold it into the batter. Pour the batter into a measuring jug.

- Heat the waffle iron, brush it with melted butter, heat it again, then fill one side with enough waffle batter to reach the top without overflowing. Cook the waffles for around half a minute or until golden-brown on both sides. Remove the waffles and serve warm with maple syrup, corn syrup, jam, butter, or ice cream.

- Use the remaining batter to make more waffles, each time brushing any debris from the waffle iron and re-greasing it.

# 3

# Balancing Your Diet

An acid-producing diet is the norm in the Western world. Such a diet lacks enough alkali-producing vegetables and fruit to balance the intake of acid-producing meat, fish, eggs, and cereal-grain food, and the result is chronic low-grade metabolic acidosis. Bicarbonate plays a very important part in the regulating systems that help our bodies counter acidosis.

# Acidic foods

An acidic food is *not* the same as an acid-producing food. Acidic foods taste acidic because they contain organic food acids (such as citric, acetic, oxalic, malic, pyruvic, and acetylsalicylic acids); "organic" here meaning that they contain only carbon, oxygen, and hydrogen. Some acidic foods, such as citrus fruit, vinegar, yogurt, sour cream, buttermilk, and soured milk, actually taste acidic. Others, such as beer, apple cider, honey, molasses, maple syrup, coffee, and "natural" cocoa and chocolate (dark unsweetened varieties made from cocoa beans that have not been "Dutch-processed") are acidic but don't taste so.

Organic food acids don't produce acid in the body because the liver breaks them down into carbon dioxide and water. Excess carbon dioxide is exhaled and excess water is eliminated by the kidneys and lungs. So most acidic foods have virtually no effect on our acid–alkali balance; a few, including lemon juice and apple cider vinegar, even have an alkali-producing effect.

# Acid-producing foods

These foods are not acidic and do not taste acidic, but when digested and metabolized produce acids that make body fluids less alkaline. An acid-producing diet is rich in meat, grain and sugar, and carbonated drinks, and low in vegetables and fruit.

Certain acid-producing foods produce acid by releasing more strong acidic anions (such as sulfate, which can join with hydrogen ions to form sulfuric acid) than strong alkaline cations (see page 97), such as potassium, which can join with bicarbonate to form potassium bicarbonate.

The kidneys can eliminate only a limited load of strong acidic ions each hour, so any surplus causes acidosis.

Acid-producing foods include:
- Protein (such as in meat, fish, eggs, cheese, pulses, or dry beans)
- Grain foods (such as bread, breakfast cereal, pasta, cakes, cookies, many puddings, and rice. Note that refined-grain foods are more acid-producing than wholegrain foods)
- Sugar (especially refined sugar—both white and brown).

Certain acid-producing foods and food constituents have an acidifying effect that is independent of the diet's total acid-producing load:
- Table salt (sodium chloride) encourages acidosis by decreasing the blood's strong-ion difference. (It does this by increasing chloride more than sodium and encouraging the kidneys to excrete potassium.)
- Tea, coffee, cocoa, chocolate, and legumes (such as soy beans and chickpeas) contain purines that are metabolized to uric acid.
- Alcohol encourages acidosis by increasing lactate production.

### FOODS THAT CAUSE AN "ACID TUMMY"

Strongly acidic foods (such as lemon juice), plus certain acid-producing foods and drinks, can stimulate the production of enough stomach acid to cause "acid indigestion." Such foods include those made from refined sugar or refined flour; tea, coffee (even decaffeinated), and cocoa; and beer and wine (possibly because of protein-breakdown products such as amino acids and amines produced during fermentation).

# Alkali-producing foods

These foods contain more strong alkaline than strong acidic ions, so when metabolized, they produce alkaline salts such as bicarbonate. They include: vegetables and fruit, certain nuts, beans, grains, cheeses, and sugars.

Many experts assert that alkali-producing foods should compose 80 percent of our diet, with acid-producing foods making up the rest. However, the average Western diet is composed mainly of acid-producing foods, with vegetables and fruit (the major alkali-producing foods) contributing only 2 percent!

# Our diet

The effect of our overall diet on our acid–alkali balance is more important than the effects of individual foods. This is because, when a meal has been digested and metabolized, its net effect is either acidic or alkaline.

Our aim should be to eat an alkali-producing diet. We can deal with temporary acidosis from an occasional acid-producing meal, but weeks or months of chronic low-grade acidosis may cause symptoms.

Research published in the *American Journal of Clinical Nutrition* (1998; 68:576–583) suggests the two key factors affecting our acid–alkali balance are our intake of the following:

- protein (especially meat), which is acidifying

- potassium (most importantly, from vegetables and fruits), which is alkalinizing.

The US Third National Health and Nutrition Examination Survey of 33,994 people in 1988–94 found that the average American diet is acid-producing. However, studies suggest that in the pre-agricultural societies of over 70,000 years ago, 87 percent of people ate an alkali-producing diet. So it is likely that a huge change in eating patterns has occurred over the millennia.

## Protein

Animal protein is found in meat, fish, eggs, and dairy food. Research suggests that dairy food's high calcium levels help to protect against any acidosis-promoting effect of its protein. Vegetable protein is found mainly in dry beans and peas, and in nuts, grains, and other seeds.

## Potassium

Potassium is plentiful in vegetables and fruits and produces strong alkalizing ions in body fluids. Particularly rich sources include green leafy vegetables, tomatoes, bananas, dates, and avocados.

### WHAT ARE ACID-PRODUCING AND ALKALI-PRODUCING FOODS AND DRINKS?

Burning a food in a laboratory produces heat and reduces the food to acidic or alkaline ash. This is the equivalent of our cells burning food to produce energy. So measuring the pH of a burnt food's ash indicates the acid- or alkali-producing effect of that food in our body. The ash produced by the average vegetarian diet is significantly more alkaline than that from the average omnivorous diet. This is mainly because animal protein has a particularly high acid-producing potential.

## THE ACID-PRODUCING POTENTIAL OF VARIOUS FOODS

"High," "Medium," and "Low" refer to a food's acid-producing potential. Most of us could do with eating less "high" acid-producing food:

| Meat and fish | High: | Pork, beef, shellfish |
|---|---|---|
| | Medium: | Eggs, lamb, sea fish, chicken |
| | Low: | Freshwater fish |
| Dairy foods | High: | Hard cheese, ice cream |
| | Medium: | Soft cheese, cream |
| | Low: | Yogurt, cottage cheese, milk |
| Grain foods | High: | White flour, white bread, white pasta |
| | Medium: | Wholemeal and wholegrain bread; cookies, white rice, corn, oats |
| | Low: | Brown rice, sprouted-wheat (Essene) bread, spelt flour and bread |
| Nuts | Medium: | Pistachios, peanuts, cashews, walnuts |
| | Low: | Macadamias, hazelnuts |

| Fats and oils | Low: | Margarine, sunflower oil, corn oil, butter |
|---|---|---|
| Sugar, honey, confectionery | Medium: | Chocolate, sugar, brown sugar |
| | Low: | Processed honey, molasses |
| Condiments | High: | Most vinegars, soy sauce, salt |
| | Medium: | Mustard, mayonnaise, tomato ketchup |
| Drinks | High: | Spirits, beer, soft drinks |
| | Medium: | Coffee, wine, fruit juice |
| | Low: | Tea |
| Vegetables | Medium: | Potatoes without skins; pinto, navy, and lima beans |
| | Low: | Kidney beans |
| Fruits | Medium: | Cranberries |
| | Low: | Plums, prunes |
| Other | Low: | Coconut milk |

## THE ALKALI-PRODUCING POTENTIAL OF VARIOUS FOODS

"High," "Medium," and "Low" refer to a food's alkali-producing potential. Most of us could do with eating or drinking more of all these foods:

| Fruits | Medium: | Avocado, tomato, lemon, dried figs, rhubarb |
|---|---|---|
| | Low: | Pineapple, raisins, dried dates, strawberries, grapefruit, apricot, blackberries, orange, peach, raspberries, banana, grapes, pear, blueberries, apple, coconut |
| Vegetables | High: | Cucumber, sprouted seeds |
| | Medium: | Radishes, celery, garlic, spinach, beets, string beans, carrots, chives, turnips |
| | Low: | Watercress, leeks, zucchini, peas, cabbage, cauliflower, mushrooms, rutabaga, onion, lettuce, potatoes with skins, asparagus, Brussels sprouts, sweet potatoes |
| Beans | Medium: | Navy beans, fresh (or frozen) soy beans |
| | Low: | Tofu, soy flour, lentils |
| Dairy food | Low: | Goat cheese, goat milk, soy milk, buttermilk |
| Grains | Low: | Buckwheat, spelt, wild rice, quinoa |
| Nuts and seeds | Low: | Almonds, Brazil nuts, chestnuts, pumpkin seeds, sunflower seeds, flaxseeds, sesame seeds |

| Oils | Medium: | Flaxseed oil |
| | Low: | Canola oil, olive oil |
| Sugar, honey | High: | Raw honey, raw sugar |
| | Medium: | Maple syrup |
| Drinks | High: | Herb tea, lemon water |
| | Medium: | Green tea |
| | Low: | Ginger tea |
| Other | Low: | Apple cider vinegar |

# How to improve your diet

If you have been eating a typical Western diet, help to rebalance it by using the above lists as a guide to favoring alkali-producing foods. An added bonus is that the vegetable and fruit content of such a diet provides a wealth of health-enhancing factors, including vitamins, phenolic compounds, carotenoids, plant hormones, salicylates, fiber, and omega-3 fats.

- As already noted, research shows that the two factors that best predict acidosis are too much protein (especially too much red meat) and not enough potassium (mainly from vegetables and fruit). However, although the matter is still being studied, my understanding is that a lack of vegetables and fruit is the most likely cause of health problems from a typical Western diet.

- Eat at least five servings a day of vegetables and fruit, of which three should be vegetables, two fruit. Potatoes (and yams, cassava, and plantain) are important for an alkali-producing diet as they are potassium-rich. However, they don't count toward your five-a-day as they are classed not as vegetables but as starchy carbohydrates (like bread, rice, noodles, pasta, and sweetcorn). Peas, beans, and lentils count as only one helping a day, no matter how much you eat, because they contain fewer nutrients than do other vegetables. However, sweet potatoes, parsnips, rutabagas, and turnips *do* count towards your five-a-day. Fruit and vegetable juices count as only one helping a day, however much you drink.

- Five-a-day is officially recommended in the US and many other countries. But one large US study (*American Journal of Preventive Medicine*, 2007) found that only 32 percent of people met the guidelines for vegetables, only 28 percent for fruits, and fewer than 11 percent for vegetables *and* fruit. Also, one in four people ate no vegetables at all on a daily basis and three in five ate no whole fruits!

- Some experts recommend up to nine helpings of vegetables a day, as well as two of fruit. This would certainly decrease your appetite for acid-producing foods!

- Cook vegetables lightly and try to use the potassium-rich cooking water in soups or gravy, or eat vegetables raw.

- Include alkali-producing food in every meal and snack.

- Keep a food diary of everything you consume over seven days. Highlight acid-producing foods in red, alkali-producing in green. This will help you to evaluate the balance of your diet.

- Use the "New American Plate" or "MyPlate" guide to eating as recommended by the US Department of Agriculture (see page 181).

- Avoid added salt; use less, or choose potassium-enriched "low" salt.

# 4

# Your Body's Bicarbonate

Sodium bicarbonate is an essential alkaline component within the body that acts as a buffer to maintain normal levels of acidity in the blood and other body fluids.

Good health (and, indeed, life itself) relies on our blood and other body fluids having the right acid–alkali balance. The bicarbonate that is naturally present in our bodies helps this to happen.

There is a lot of scientific information in this chapter, but, as the poet T. S. Eliot said, "Wisdom may be lost in information," so if you prefer to skip the science, the practical ways of correcting an acid–alkali imbalance are given on pages 78–87. Note that if do you decide to read this chapter, there is a glossary explaining many of the technical terms at the end (see page 111).

# Where does the bicarbonate in your body come from?

The building blocks of your body's bicarbonate originate from food and drink:

- As our cells produce energy from sugar (which is mainly derived from dietary carbohydrate), they release carbon dioxide. This can react reversibly with water to form carbonic acid—a weak acid that then breaks down reversibly into bicarbonate and hydrogen.

- Many foods contain carbonates or other organic substances that the body can convert into bicarbonate. ("Organic" is used in its chemical sense, meaning the substances contain only carbon, oxygen, and perhaps, hydrogen.)

- Any bicarbonate consumed in food or drink neutralizes a certain amount of stomach acid; any excess is absorbed from the intestine into the blood.

- Some people also get bicarbonate from antacids or other medications.

# Know your body fluids

Fluid accounts for nearly 60 percent of the weight of a man's body and 55 percent of a woman's. The average man's body, weighing 154lb (70kg), contains 89 US pints (42 liters), comprising:

- 53 US pints (25 liters) of intracellular fluid (fluid in cells).

- 36 US pints (17 litres) of extracellular fluid (fluid outside cells), including:

  - 24 US pints (11.5 liters) of tissue fluid (fluid that bathes cells)

  - 10.5 US pints (5 liters) of blood (circulating in blood vessels)

  - 2 cups (500 ml) of other extracellular fluids (including lymph bile, saliva, pancreatic juice, spinal fluid, joint fluid, and eye fluid).

---

### DID YOU KNOW?

Water is the basis of each body fluid. A lack of water—known as dehydration—is particularly likely in elderly people, because their perception of thirst is less acute, and their kidneys are less efficient. The possible effects include slow mental function, constipation, headaches, infections, kidney stones, heart attacks, and strokes.

# Acid–alkali balance

Each body fluid is a watery solution containing electrolytes (substances that can break down into electrically charged particles called ions). A fluid is said to be *acidic* if its concentration of hydrogen ions is greater than that of its hydroxyl ions, and *alkaline* if the other way around.

One hydrogen ion and one hydroxide ion can combine to form a molecule of water. A solution is *neutral* (neither acidic nor alkaline) if its hydrogen and hydroxide-ion concentrations are equal.

# The body's acids include:

- **Carbonic acid.** This is called a "volatile" acid because it can break down into carbon dioxide and water, which can be breathed out by the lungs.

- **Strong inorganic acidic anions**, including chloride (from the breakdown of dietary proteins) and sulfate (made from dietary proteins by the breakdown of the sulfur-containing amino acids methionine and cysteine).

- **Weak inorganic acidic anions,** mainly phosphate (from the breakdown of dietary proteins) and albumin (a blood protein made by the liver) but also nitrate and the amino acid glutamine (from dietary proteins), and other blood proteins.

- **Organic acidic anions**, including lactate, pyruvate, malate, formate, and ketones, made by our cells.

## WHAT IS PH AND WHY IS IT IMPORTANT?

A body fluid's pH ("potential of Hydrogen") represents its hydrogen-ion concentration, which in turn indicates its acid–alkali balance. Each number on the pH scale represents a ten-fold decrease in hydrogen-ion concentration from the one below and a ten-fold increase from the one above.

- A fluid becomes more acidic as its hydrogen-ion concentration rises and its pH falls, and becomes more alkaline as its hydrogen-ion concentration falls and its pH rises.

- The pH scale goes from 0 to 14, with 7 neutral for water. Because pH is temperature-dependent, a neutral pH for blood at body temperature would be 6.7. But we would die if it were neutral! It must always be alkaline. And it must stay within a very small range of pH.

# The pH of body fluids

Most body fluids are alkaline. Their pHs are regulated by our lungs, kidneys, and buffers (see page 98). And they affect each other in various important ways.

- Normal arterial blood and tissue fluid have an average alkaline pH of 7.4 (range 7.35–7.45). Venous blood is slightly less alkaline at 7.36 (since cells produce acids, releasing hydrogen ions that can enter veins).

- Intracellular fluid has a pH of about 7.

- Pancreatic juice is particularly alkaline at pH 7.5–8.8. Urine can be acidic or alkaline (pH 4.5–7.5). Stomach juice is very acidic (pH 1–2). As stomach contents enter the duodenum (the first part of the small intestine), they are alkalinized by bile and pancreatic juice. The skin's surface moisture, from skin oil and sweat, mostly has a pH of 4.5–5.75, but is slightly less acidic in the armpits and around the genitals.

# Does a change in pH matter?

A change in the pHs of blood, tissue fluid or intracellular fluid can have vitally important consequences to our health, well-being and even life itself, by affecting:

- Energy production (measurable as a person's "basal metabolic rate")

- Molecular reactivity

- Oxidation (a low pH encourages the production of "free radicals"—also known as reactive oxygen-containing ions)

- Bonding of oxygen and carbon dioxide to hemoglobin (the pigment in red blood cells) and therefore the transport of oxygen to cells and carbon dioxide from cells

- Muscle-cell contraction

- Enzyme activity

- "Folding," and therefore function, of proteins (including structural ones)

- Exchange of potassium and sodium across cell membranes

- Bio-electric signaling in and between cells

- Calcium balance

- Fatty acid and cholesterol metabolism

- Levels and activity of, and sensitivity to, certain hormones (for example, adrenaline, thyroxine and growth hormone)

- Cell growth, differentiation, multiplication, and apoptosis ("cell suicide")

- Cell mobility

- Behavior of red blood cells (a low pH makes them stack up so they cannot properly transport oxygen, carbon dioxide, nutrients, and waste products)

- Size and behavior of white cells (a low pH makes them smaller and much less active)

The degree of pH change required to affect cells isn't always clear. However, a very abnormal pH (below 6.8 or above 7.8) makes the normal folded shape of each protein molecule begin to "unravel." If this continues, life can continue for only a few hours.

## How the body responds to changes in pH

When the body-fluids' normal pHs are threatened or actually changed by diet, lifestyle, disease, or medication, the following things happen:

- Lungs exhale more carbon dioxide to reduce acidity, or less to increase acidity.

- Kidneys excrete more strong inorganic acidic anions (mainly chloride) to reduce acidity, or fewer to increase acidity. This pH-regulation is slower than that of the lungs.

- Body fluids buffer (minimize) pH changes by changing the levels of bicarbonate, non-volatile weak inorganic acidic ions or acids.

- Liver can metabolize organic acidic anions (for example, it converts lactate into glycogen).

- Cell membranes allow a greater or smaller passage of various ions between cells, tissue fluid, and blood.

- Sympathetic nervous system produces more or less adrenaline to stimulate the heart, lungs, and kidneys.

- Sweat contains more or fewer acidic anions (such as phosphates, sulfates, chlorides, and lactates).

- The above responses affect the following independent pH-regulating factors in body fluids:

  1 Strong-ion difference. Most body fluids are alkaline because their concentration of strong alkaline cations (sodium, potassium and, to a lesser extent, calcium, and magnesium) is greater than that of strong acidic anions (chloride, sulfate, phosphate, lactate). The strong-ion difference and, therefore, the pH, fall with a decrease in strong cations or an increase in strong anions, and vice versa. The strong-ion difference is affected by our diet and the activity of our digestive tract, kidneys, and cells.

  2 Carbon dioxide. This flows from cells to tissue fluid to blood. It can also flow the other way. The amount in our body fluids varies with changes in metabolism, circulation and breathing rate.

  3 Non-volatile weak inorganic acids.

Changes in the above factors in turn change our bicarbonate and hydrogen-ion levels.

## Bicarbonate and the lungs

Carbon dioxide produced by cell metabolism is carried in the blood to the lungs, where any excess is breathed out to prevent it reducing our blood pH (and so making it less alkaline).

If something threatens to lower our blood pH, we automatically breathe more rapidly and exhale more carbon dioxide, and vice versa.

## Bicarbonate and the kidneys

Our kidneys filter bicarbonate from the blood and excrete about 20 percent of it. If something threatens to lower our blood pH, the kidneys excrete less bicarbonate, and vice versa.

## Bicarbonate buffer

Buffers change their concentration to help maintain the body's acid–alkali balance. Bicarbonate is our most important buffer. Bicarbonate ions can raise pH (thereby increasing alkalinity) by joining with hydrogen ions to form carbonic acid.

Carbonic acid can lower pH (thereby reducing alkalinity) by breaking down into hydrogen and bicarbonate ions. Carbonic acid is such a weak acid that in itself it has negligible acidity.

# Acidosis and alkalosis

Acidosis is a "push" toward acidity, or, more accurately (because blood, tissue fluid, and intracellular fluid are never actually acidic), a push toward lowered alkalinity. This is associated with the accumulation of acid and hydrogen ions, or loss of bicarbonate.

The lungs, kidneys, and buffers try to compensate so as to maintain or restore a normal blood pH. While this is perfectly normal, if prolonged or extreme it can "strain" the body and cause symptoms. Depending on the success of compensation, the blood pH either remains within its normal range or falls below 7.35 (which causes symptoms—see below).

Alkalosis is a "push" toward over-alkalinity. The lungs, kidneys, and buffers try to compensate for this. Depending on their success, the blood pH may remain within its normal range or rise above 7.45. Alkalosis is less common than acidosis.

# Lactic acidosis

However, acidosis isn't all bad. For example, lactic acidosis associated with lactate production in muscle cells during strenuous exercise increases the oxygen available to them.

Certain people's cells oxidize sugar (to produce energy) particularly rapidly. This encourages lactic acidosis. Such people are known as "fast oxidizers" and tend to feel hungry after a meal sooner than do others.

**Note that:**

- Symptoms from acidosis or alkalosis result from changes in blood, tissue fluid, and intracellular fluid of the levels of strong ions (such as sodium, potassium, chloride, and sulfate); the amount of carbon dioxide; and the concentration of non-volatile weak inorganic acids. Changes in these independent pH-regulating factors are accompanied by changes in levels of hydrogen and bicarbonate ions.

- It is vital to try to identify and treat the underlying cause.

- The causes of acidosis and alkalosis are metabolic, respiratory, or mixed.

The lists here and on the following pages are based on acidosis below pH 7.35 ("acidemia") and alkalosis above pH 7.45 ("alkalemia"). But certain causes, symptoms, and treatments also pertain to less severe acidosis or alkalosis.

## Metabolic acidosis

This results from a build-up of acids from an acid-forming diet, increased acid production, reduced acid excretion, or loss of alkali. There may also be too much chloride. The causes include:

1 Lactic acidosis from increased or impaired energy production in cells, resulting from:

- strenuous exercise

- severe infection

- prolonged or widespread inflammation

- prolonged lack of oxygen, for example, from a seizure, severe anemia, shock, lung disease, heart failure

- low blood sugar, which increases adrenaline (so as to increase blood sugar) and thereby speeds muscle-cell metabolism

- too much alcohol, especially if a person lacks vitamin B1

- prolonged stress, anxiety, or anger, as this increases adrenaline

- smoking

- certain drugs including metformin (for diabetes)

- liver failure

- certain cancers (such as lymphoma and leukemia).

2 Uncontrolled diabetes, a low-carbohydrate diet, or starvation, which means that fats are used instead of sugar for energy, producing acidic substances called ketones.

3 Aspirin and certain other medications.

4 Severe diarrhea originating in the small intestine and causing a loss of bicarbonate, or the genetic kidney disease proximal renal tubular acidosis, which causes a loss of bicarbonate in urine or the retention of hydrogen ions, or both.

5 Kidney failure or distal renal tubular acidosis, which reduce urinary excretion of acidic sulfates, phosphates and urates.

6 An acid-forming diet and an age-related decline in kidney function, though these cannot in themselves reduce the pH below 7.3.

*What happens?* The lungs try to compensate with deep rapid breathing that reduces the blood's level of carbon dioxide. The kidneys try to compensate by excreting more hydrogen and less bicarbonate.

*Symptoms:* headache, anxiety, drowsiness, fatigue, poor vision, nausea, vomiting, abdominal pain, altered appetite, muscle weakness, bone pain.

Extreme acidosis can cause weakness, confusion, convulsions, heartbeat abnormalities, and low blood pressure.

Lactic acidosis makes breath smell of acetone (nail-polish remover) or pear drops. People with coronary artery disease may have an abnormal heartbeat and angina (chest pain on exercise).

*Treatment:* Identify and treat the cause. For very severe acidosis (below pH 7.1), doctors may give intravenous (IV) sodium bicarbonate. If less severe (pH 7-7.35), they occasionally give IV or oral sodium bicarbonate. Oral bicarbonate can help soon after an aspirin overdose as it speeds up the elimination of aspirin in the urine.

## Respiratory acidosis

This is associated with excess retention of carbon dioxide in the blood, resulting from abnormally slow or interrupted breathing. The causes include:

- Certain medications, including anesthetics and sedatives

- Brain injury, tumor, or infection affecting the brain's breathing-control center

- Playing a wind instrument, swimming underwater, or other activity involving spasmodic breathing

- Pneumonia or other problems that reduce the exchange of oxygen and carbon dioxide in the lungs

- Severe asthma, chronic bronchitis, emphysema, pneumonia, pneumothorax, airway obstruction, or other problems that increase airway resistance

- Obesity, burns, chest injury, polio, muscular dystrophy, Guillain-Barré syndrome, chest-wall deformity, scoliosis, or other conditions that impede breathing by reducing chest-wall movement.

*What happens?* The kidneys try to compensate by excreting more hydrogen and less bicarbonate.

*Symptoms*: headache, drowsiness, fatigue, nausea, vomiting, and exhaustion.

*Treatment*: Identify and treat the cause.

## Metabolic alkalosis

This is associated with raised bicarbonate caused by the body losing too much acid or gaining too much alkali. The causes include:

- Loss of chloride due to repeated vomiting or severe diarrhea originating in the large intestine (for example, from ulcerative colitis or the overuse of laxatives), or certain diuretics (medications that increase urine production)

- Low potassium from a poor diet, vomiting, overactive adrenal glands, or diuretics. The kidneys retain sodium to take its place, but the resulting loss of hydrogen in urine makes blood over-alkaline

- Dehydration, diuretics, heart failure, bleeding, severe burns, or other causes of low blood and extracellular-fluid volume. The kidneys respond by retaining sodium, but the resulting hydrogen loss makes blood over-alkaline

- Steroid medication

- Too much IV bicarbonate medication

- Rarely, too much antacid medication.

*What happens?* Compensation by the body's pH-balancing mechanisms returns the pH to normal but leaves bicarbonate and carbon dioxide high.

*Symptoms:* Nausea and vomiting; numbness or tingling in the hands, feet, and face; tremor; muscle twitching; cramp; dizziness; fainting; confusion.

*Treatment:* Identify and treat the cause. Water is replaced, if necessary with sodium and potassium, by IV infusion. Very severe alkalemia (with blood pH higher than 7.6) may be treated with IV dilute acid.

# Respiratory alkalosis

This is associated with low carbon dioxide in the blood, caused by rapid breathing. The causes include:

- Anxiety

- Pain

- Fever

- Aspirin overdose (can also cause metabolic acidosis)

- Low oxygen from lung disease or high altitude.

*What happens?* The kidneys try to compensate, partly by excreting more bicarbonate.

*Symptoms*: Nausea and vomiting; numbness or tingling in the hands, feet and face; tremor; muscle twitching; cramp; dizziness; fainting; confusion.

*Treatment*: Identify and treat the cause. Slower breathing can help if anxiety or pain is responsible, as can breathing in and out of a large paper (not plastic) bag to raise the blood's carbon-dioxide level.

## Does mild acidosis cause symptoms?

Researchers claim that almost all adults eating a typical Western diet have mild ongoing acidosis. The question is whether this "chronic low-grade metabolic acidosis" can cause symptoms. Experts have long contested the idea, but we now know it can happen. For an excellent review, read, "Diet-induced acidosis: is it real and clinically relevant?" in the *British Journal of Nutrition* (2010) 103, 1185–1194.

Whether such symptoms result from the lungs, kidneys, and buffers compensating for the threat to the normal range of pH, or from a slight reduction in pH that nevertheless remains within the normal range, is unclear.

## What causes mild acidosis?

It seems sensible to assume that any cause of metabolic acidosis with a pH below 7.35 could, if less severe, cause mild metabolic acidosis. But researchers think the two *main* causes of mild metabolic acidosis are:

- An acid-forming diet

- An age-related decline in kidney function.

We cannot prevent aging, but we can adjust our diet (see pages 78–87). We can also deal with conditions that can cause more severe metabolic acidosis and might, therefore, cause the mild type too. Finally, we can consider taking an alkali supplement.

# Acid–alkali balance in the digestive tract

The efficiency of saliva, stomach juice, bile, pancreatic juice, and intestinal juice depend on their pH.

## Saliva

A high level of bicarbonate in saliva helps to neutralize acidic foods and enable the pytalin (an enzyme also called salivary amylase) to start breaking down starches into sugar. The pH of saliva between meals is 6.3 or above.

## Stomach

Certain stomach-lining glands produce mucus that has a pH of 7.7 and contains large quantities (up to 0.5g a day) of sodium bicarbonate. This alkaline mucus protects the stomach from its own acid. Bicarbonate production rises if blood pH falls, and vice versa.

Other stomach-lining glands enable sodium and chloride ions to join with water and carbon dioxide to form hydrochloric acid (stomach acid) and sodium bicarbonate. For each molecule of acid produced, one of sodium bicarbonate is produced, too. All food stimulates stomach-acid production by distending the stomach, but protein is the main stimulus. Hydrochloric acid is released into the stomach, but bicarbonate ions are absorbed from the glands into the blood. Hydrochloric acid (stomach acid) has a pH of 1–3 and, along with the enzyme pepsinogen, alters proteins to make them more easily digestible. It also kills most of the micro-organisms we consume.

**Alkaline tide** Blood leaving the stomach is relatively alkaline as it is rich in bicarbonate and short of chloride. As this "alkaline tide" circulates around the body, its raised alkalinity helps to reduce any acidosis resulting from a meal.

The more acid-producing a meal (meaning the more acid it produces after digestion and metabolism), the more sodium bicarbonate is produced by the stomach and absorbed into the blood. That way the body fluids have enough to buffer a meal's acid-forming effects. However, the more sodium bicarbonate that is produced, the more stomach acid is produced, too. So an acid-forming meal increases stomach-acid production. Any excess acid is absorbed into the blood, which adds to its acid load. Conversely, when we have an alkali-producing meal, the stomach doesn't need to produce large amounts of sodium bicarbonate, so it doesn't produce large amounts of stomach acid either.

Several hours after a meal, the stomach contents enter the duodenum as an acidic "sludge."

**Low stomach-acid production** This can result from aging, stress, the prolonged use of antacid or acid-suppressant medication, or an acid-producing diet. It affects one in two people aged over 60.

It is possible, though unproven, that low stomach-acid production might also result from insufficient dietary potassium (mainly from vegetables and fruit). The body would then use sodium instead of potassium to accompany acids as they are excreted in urine. Sodium would be mobilized from various places, including stomach-acid producing cells, and this would reduce stomach-acid production. Possible problems from too little stomach acid include:

- Malnutrition—because a lack of acid discourages the absorption of proteins, vitamins B and C, calcium, chromium, iron, magnesium, selenium, and zinc

- Gallstones—because too little acid discourages gallbladder contractions, and gallstones form more easily in stagnant bile

- Allergy—because too little acid means proteins aren't prepared for digestion; poorly digested proteins can then be absorbed from the intestine into the blood and cause allergic sensitization

- Gastritis and peptic ulcer—because too little acid strongly correlates with *Helicobacter pylori* bacterial infection. An ulcer can develop if anything interferes with the stomach lining's cells or protective mucus. The usual culprit is inflammation (gastritis) from *H. pylori* infection. This also encourages stomach cancer.

# Liver
Bicarbonate in bile helps to alkalinize stomach contents after their arrival in the duodenum. The liver also produces enzymes that alkalinize blood. Raised levels (indicated by liver-function blood tests) suggest that the liver is working overtime to help maintain a normal blood pH. The liver also influences the concentration of weak acids in body fluids.

# Pancreas
Pancreatic juice is very alkaline because it receives a lot of sodium but not much chloride from the blood.

Secretin, a hormone produced by the duodenum, stimulates bicarbonate production in pancreatic-duct cells. The more acidic the food residues in the duodenum, the more secretin is produced.

# Small intestine
Many factors influence the pH in the duodenum, including the size and timing of meals, the proportions of strong cations and anions (see page 97) from food,

and the volume and pH of the stomach contents, bile, and pancreatic juice. The resulting pH of about 7 activates pancreatic enzymes, including trypsinogen and chymotrypsinogen (which break down proteins into amino acids), lipase (which breaks down fats into fatty acids) and amylase (which breaks down starches to sugar).

Glands in the lining of the small intestine produce large amounts of alkaline fluid containing bicarbonate, which helps to neutralize any remaining stomach acid. Chloride from stomach acid, bicarbonate from pancreatic juice, and strong ions from food are absorbed into the blood. All this takes hours, which suits the slow pace of the kidneys' pH-regulating mechanisms.

## Colon

Water and strong ions, such as sodium and potassium, can be absorbed here. The colon is more alkaline than the small intestine because it has less chloride.

### DID YOU KNOW?

Low production of stomach acid becomes more common with age and is a possible cause of many ailments. By the age of 40, almost 40 percent of the population are affected, and this increases to over 50 percent by the age of 60.

# Glossary

**Acid:** a substance that can neutralize a base and increase the hydrogen-ion concentration of a watery solution.

**Alkali:** a base that increases the hydroxide-ion concentration of a watery solution.

**Anion:** a negatively charged ion. **Strong acidic anions:** anions that can form acids (for example, hydrochloric, sulfuric, and lactic) when combined with hydrogen ions.

**Base:** a substance that can neutralize an acid and decrease the hydrogen-ion concentration of a watery solution.

**Cation:** a positively charged ion. **Strong alkaline cations:** strong cations that can form alkali when combined with weak anions.

**Electrolyte:** a substance that reversibly breaks down into ions in a watery solution. Strong electrolytes break down into ions completely in a watery solution. Weak electrolytes break down only partially.

**Hydrogen atom:** an electron (a particle with a negative electrical charge) plus a proton (a particle with a positive charge).

**Hydrogen ion:** a hydrogen atom minus its electron. This positively charged particle is 1800 times smaller than a hydrogen atom so has a particularly powerful electric field. This makes it react unusually rapidly with other ions, explaining its importance in acid-base balance.

**Hydroxide ion:** an ion made of one atom each of oxygen and hydrogen that between them have lost a proton, producing a negative electrical charge.

**Ion:** an atom or molecule that, when dissolved in water, has a positive or negative charge because it loses or gains one or more electrons. Strong ions come from strong electrolytes. Weak ions come from weak electrolytes.

5

# Beauty and Personal Care

Cleansing, soothing, and softening, baking soda is a great aid to personal care, and can be used in a wide range of beneficial treatments for the whole body.

Baking soda is a highly economic and effective addition to anyone's beauty regime. For one thing, it softens water, so you need less soap to wash your face and body, and less shampoo to wash your hair. This is beneficial because the more soap or shampoo you use, the more you strip valuable natural oils and acidity from your skin and scalp.

Another benefit is that its very mildly abrasive nature means that it helps to cleanse the skin. Its tiny rounded particles also have gentle exfoliating properties, helping to dislodge dead cells. This smoothes and brightens the skin, making it look younger and healthier. In addition, baking soda stimulates the skin's tiny blood vessels (capillaries), giving the skin a soft attractive glow. It can also be used to help to clear blackheads.

When present in toothpaste, baking soda can help to remove dental plaque (the sticky film of food residues and bacteria that accumulates on teeth) and tartar (hard, calcium-impregnated plaque that encourages gum disease). Its alkalinity is helpful for teeth, too. For example, it can neutralize the acidity released during the breakdown of sugars and other refined carbohydrates by mouth bacteria. This is important, because acidity encourages teeth to decay.

Baking soda's alkalinity has antiseptic and antifungal actions that can help to prevent or treat minor skin and scalp infections. It also has a deodorant action that is thought to result from the neutralization of odoriferous short-chain fatty acids as found, for example, in the sweat produced in the armpits.

Finally, baking soda is a common ingredient in the effervescent balls of bath salts known as bath "bombs."

On the following pages you'll find tips on how to use baking soda as an all-round beauty aid, as well as recipes for toiletries such as toothpaste, deodorant, and bath "bombs."

# Bathing

To soften bathwater add ½ cup (3½ oz, 100 g) baking soda to your bath. A few drops of fragrant oil will make your bathtime particularly sybaritic. Cedarwood, frankincense, and lavender essential oils are said to be relaxing, while geranium, jasmine, neroli, and ylang ylang are considered uplifting, and cardamom is stimulating and refreshing.

For extra cleaning power, so you don't need to use soap, add several squeezes of shampoo to the mix.

For a seaside-spa-style bath, add ½ cup (3½ oz, 100 g) baking soda plus ½ cup (4 oz, 100 g) sea salt to the water.

To make bath "bombs" (colorful fragrant lumps of bath salts that fizz when added to bath water), mix 2 cups (14 oz, 200 g) baking soda and 1 cup (7 oz, 200 g) citric acid (from a drugstore or the internet) together in a bowl. Add 10 drops of fragrant essential oil plus a few drops of food coloring. Now add about 1 teaspoon of almond oil or baby oil, a few drops at a time, until the mixture is crumbly but just clumps together when firmly squeezed. Firmly press it into molds such as plastic egg boxes, silicone, or other nonstick cupcake molds, little yogurt pots, or special shaped molds from craft stores. Leave to dry for a week. Then, when you have a bath, put a bath bomb in the water and watch it fizz as its citric acid and baking soda react with the water to release thousands of bubbles of carbon dioxide.

# Hand protection

If you dislike wearing rubber gloves when washing dishes, help to keep your hands soft by adding 1 tablespoon of baking soda to the water before you set about tackling a basin of dirty dishes.

# Deodorizing

- Adding baking soda to your bath water (as described on the previous page) will help to keep you odor-free.

- In a small bowl, mix ½ teaspoon of baking soda, 1 or 2 drops of water and, if required, 1 drop of fragrant essential oil. Smooth a little of this deodorant paste under your arms.

- Alternatively, pat on baking soda as a dry deodorant.

- To help keep feet fresh and dry, sprinkle baking soda between your toes, or into pantyhose or socks, as an alternative to talcum powder.

- Shake a little baking soda on a new sanitary pad before use.

# Exfoliating

- After cleansing your skin, make an exfoliant paste by mixing about 3 parts of baking soda to 1 part of water. Gently rub this into your skin, then rinse off.

- To smooth roughened lips, rub them with a paste made by mixing baking soda with lemon juice. Rinse off and then apply some lip balm.

- Smooth nail cuticles by wetting them, rubbing in a little baking soda, then rinsing, drying, and applying moisturizer.

# Softening dry, rough, or hard skin

Make a skin-softening paste by combining baking soda and extra-virgin olive oil. Add fragrance, if you like, by adding a few drops of lavender, rose, or neroli essential oil. Smooth the paste over your hands, elbows, the bottoms of your feet or other areas with dry, rough, or hard skin. Wait for 30 minutes, then rinse off the paste with warm water and immediately smooth in some moisturizing cream.

# Shaving

- Mix baking soda with water to form a loose paste and apply this to your skin before shaving as an alternative to soap lather or shaving cream or gel. This is less "drying" to the skin than most soaps because it doesn't reduce the skin's natural oils and acidity.

- Quickly soothe a shaving rash by smoothing on a paste made from baking soda and water, as above. Remember to wash the paste off visible areas such as your face or legs before you see other people.

- Help to prevent a shaving rash by smoothing on a paste made from baking soda and water, as above.

## CLEARING BLACKHEADS

Twice a day, clean and dry your face. In a small bowl, mix 1 tablespoon of baking soda with about 1 teaspoon of water to make a paste. Rub this over your skin to help loosen and remove blackheads, leave for 10–20 minutes, then rinse. Now help to restore your skin's natural acidity by patting on a little apple cider vinegar.

# Hair care

Remove a build-up of products, such as hairsprays, gels, and mousses by squeezing your usual amount of shampoo into the palm of your hand and mixing in 1–2 teaspoons of baking soda. Apply this mixture to your hair, lather up, then rinse well. Repeat, then condition your hair. as usual

Try dry-cleaning greasy hair, especially if you are short of time, by applying just a pinch or two of baking soda to the hair, rubbing it through with your hands, then brushing your hair thoroughly.

If you swim frequently in chlorinated pools, always rinse your hair afterward with a solution of ½ teaspoon baking soda in 1 pint of water to remove the chemical residue that can discolor hair.

## CLEAN HAIRBRUSHES AND COMBS

Increase your hair-shine by making sure that your hairbrushes and combs are kept clean and fresh. Remove any dirt and hair-product build-up by leaving them to soak in a basin of warm water in which you have dissolved 1 teaspoon of baking soda. Remove after 30 minutes, rinse well and leave to dry.

# Dental care

Coat your toothbrush bristles with baking soda before brushing your teeth. This "dry-brushing" with bicarbonate removes plaque better than "wet-brushing." Baking soda cleans and brightens teeth. It also counteracts the acidity that is produced by mouth bacteria from traces of sugars and other refined carbohydrates, and which encourages tooth decay.

Some people advocate making a fresher-tasting paste by mixing equal amounts of baking soda and fine sea salt. Salt boosts the flow of saliva (which has antibacterial properties, so helping prevent tooth decay) and contains the alkaline mineral sodium, which helps to counteract acidity.

Make a whitening toothpaste by mixing baking soda with 3 percent hydrogen peroxide solution. Use it once a week and rinse your mouth thoroughly with water afterwards. Hydrogen peroxide is available from pharmacies.

Make a cleansing, refreshing mouthwash by mixing ½ teaspoon baking soda in ¼ glass of warm water. Take a mouthful and swish it around your mouth.

# Foot care

Soothe aching feet by adding 4 tablespoons of baking soda to a basin of warm water. Relax and soak your feet for 10–15 minutes. Add several drops of essential oil for fragrance: cedarwood, frankincense, and lavender are considered relaxing, while geranium, jasmine, neroli, and ylang ylang are considered uplifting, and cardamom stimulating and refreshing. After drying your feet thoroughly, massage in a little baking soda to smooth the skin.

# 6

# Ailments and Natural Remedies

Long recognized as a highly effective antacid, sodium bicarbonate is frequently prescribed to relieve heartburn and acid indigestion. It can also be used as a remedy for a broad range of other medical complaints. Be sure to read all the cautionary advice that I give on the following pages before trying any of the remedies.

You can treat yourself with sodium bicarbonate in three different ways. The first is by changing to an alkali-producing diet, as described on pages 78–87. Changing to an alkali-producing diet is the most important starting point for nearly all the common ailments listed in this chapter. Two other methods of treatment are taking sodium bicarbonate as a solution by mouth, or applying it to the skin as a paste or in a bath.

# How sodium bicarbonate can be taken

Sodium bicarbonate is available from supermarkets and grocery stores as "baking soda." It is **not** the same as "baking powder" (see page 25). It is available from pharmacies and drugstores as "sodium bicarbonate."

## Dose

One teaspoon of the powder contains 600 mg of sodium bicarbonate. The usual adult dose, taken 1–4 times a day, is:

**Under-60s: ½–1 teaspoon**

**Over-60s: ¼–½ teaspoon**

In the US, sodium bicarbonate is available as 325 mg and 650 mg tablets. Be aware that the US Food and Drug Administration (FDA) sets the maximum dose at 16 g per day for adults aged under 60; 8 g per day for the over-60s.

For optimal absorption into the blood, take sodium bicarbonate dissolved in water on an empty stomach. For acid indigestion, take it when you have any discomfort. Doctors sometimes give sodium bicarbonate in an intravenous (IV) drip to increase plasma alkalinity.

As sodium bicarbonate neutralizes stomach acid, it releases carbon dioxide, which causes belching. Any excess sodium bicarbonate passes into the intestine and, when absorbed into the blood, has an alkalinizing effect.

# How to make soda water

Drinking soda water is a pleasant way of taking sodium bicarbonate. Make it with a 4-cup (35-fl oz,1-liter) rechargeable soda siphon, a disposable one-shot screw-in cartridge of pressurized carbon dioxide, plus sodium bicarbonate and water.

Add ¼–½ teaspoon of sodium bicarbonate to 4 cups (35 fl oz, 1 liter) of water. Put this into the soda siphon, then add the carbon dioxide.

Flavor it with plain syrup or fruit syrup if you like.

Unlike many soft drinks, soda water does not contain phosphoric acid, which encourages calcium loss from bones.

# Alternatives to sodium bicarbonate

If you cannot take sodium bicarbonate for some reason, discuss with your doctor whether an alternative alkalinizer (such as potassium bicarbonate or calcium carbonate) would be appropriate.

# Tips and warnings

- Don't take sodium bicarbonate daily for more than 2 weeks without consulting a doctor.

- Don't confuse sodium bicarbonate with baking powder.

- Unless agreed with your doctor, don't take sodium bicarbonate if you're on a sodium-restricted diet, or have high blood pressure, kidney, lung, heart or liver disease, fluid retention, urination problems, low blood calcium, or anal bleeding.

- Take only small doses if pregnant, breastfeeding, or over 60.

- Consult your doctor first if you are on prescription drugs, as some interact with sodium bicarbonate. Some of those that do interact include benzodiazepines, calcium- or citrate-containing preparations, ephedrine, fluoroquinolones, iron, ketoconazole, lithium, methenamine, oral anti-diabetes drugs, quinidine, steroids, tetracycline, and urine-acidifying medications.

- Check with a doctor or sports coach if considering taking sodium bicarbonate to enhance your exercise performance.

- Don't give sodium bicarbonate to a child unless you have first discussed this with a pediatrician and agreed the dose.

If you suspect a serious reaction, or have taken an overdose, call a doctor or contact your local poison unit or emergency room immediately, or call the National Poison Hotline on 1-800-222-1222. If you go to a doctor or hospital, you or someone else should bring the container of sodium bicarbonate.

# Be aware of possible side effects

Depending on the dose and the person, sodium bicarbonate may cause some severe side effects:

- Metabolic alkalosis, with nausea, vomiting, numbness, tingling, tremor, muscle twitching, cramp, dizziness, fainting, confusion

- High blood sodium, possibly with thirst, ankle swelling, high blood pressure, frequent urination, seizures, heart failure, or even a stroke

- Low blood potassium, possibly with weakness, fatigue, cramp, tingling, numbness, nausea, vomiting, bloating, constipation, irregular heartbeat, large amounts of urine, thirst, confusion, hallucinations, or fainting

- Nervous-system depression, possibly with headache, drowsiness, nausea, vomiting, lack of coordination, dizziness, or confusion

- Milk-alkali syndrome—abnormally high blood calcium in people with poor kidney function, if sodium bicarbonate is taken with dairy products, calcium supplements, or calcium-containing antacids. If continued, this can cause calcium deposits, kidney stones, and kidney failure

- Allergic swelling of the face, lips, tongue and throat, and difficulty in breathing. This needs urgent medical attention.

# Apply sodium bicarbonate to the skin

Sodium bicarbonate dissolved in water has an alkaline pH of 8.3 and can ease the itching, discomfort, and inflammation of certain skin conditions. Two ways of applying it are by making it into a paste, or adding it to a bath.

## Sodium-bicarbonate paste

Put 1 tablespoon of sodium bicarbonate into a small bowl and stir in about 1 teaspoon of water. Apply a thin layer of the paste to the skin. Let the paste dry on the skin, then leave it for 30 minutes before rinsing off with water.

## Alkaline bath

Add ½–1 cup (3½–7 oz, 100–200 g) of sodium bicarbonate to a bath of comfortably hot water. Add a few drops of lavender or other essential oil for fragrance, climb in and enjoy a soak.

# Testing urine pH

Unusually acidic urine suggests that the kidneys are eliminating excess acid. The most common reason for this is an acid-forming diet.

The pH of normal urine ranges from 4.6 to 8.0. It is more acidic in the morning than in the evening. If you test your urine, it's best to test "24-hour" urine (all the urine passed in 24 hours and pooled in one container, or cupfuls collected each time you urinate and pooled in one container). Urine pH test strips are available from pharmacies or on the internet.

# Ailments A–Z

Please note:

- Every ailment has many possible causes. Only those pertaining to sodium bicarbonate or the body's acid–alkali balance are included here.

- Consult a doctor about possible causes and treatments for continuing, worrying, or worsening symptoms.

- You can help prevent and treat most ailments with a healthy diet, adequate hydration, regular exercise, daily outdoor light, effective stress management, a sensible alcohol intake, and no smoking.

- When I mention a study, I give the journal's name and year of publication. These, plus some keywords, should enable you to find out more on the internet.

# Anxiety

Anxiety can encourage either acidosis or alkalosis. In addition, acidosis can encourage anxiety. One way in which anxiety can cause acidosis is by increasing the stress hormone adrenaline. This tenses muscles, which in turn boosts energy production in muscle-cells, releasing an acidic anion called lactate.

> Urine from people having a panic attack is unusually acidic.
> *(Psychiatry Research,* 2005*)*

> Increased adrenaline production can cause acidosis.
> *(Clinical Science,* 1983*)*

Conversely, some anxious people breathe rapidly, so they exhale too much carbon dioxide, encouraging respiratory alkalosis (see page 105).

As for acidosis encouraging anxiety, people on a high-protein diet have acidosis and are allegedly more likely to feel irritable. Also, chronic acidosis encourages anxiety by encouraging magnesium loss from bones.

• **Action:**
**Eat an alkali-producing diet (see pages 84–87).**

# Arthritis

Rheumatoid arthritis is linked with the intake of meat, wheat, sugar, salt, and coffee—all of which are acid-producing:

Daily meat consumption was associated with double the risk compared with eating it less often. More protein in general was implicated too.

*(Arthritis and Rheumatism, 2004)*

A study at the University of Oslo associated meat, wine, and coffee with joint swelling.

*(Plant Foods for Human Nutrition, 1993)*

Other research indicates a lower likelihood in Mediterranean countries, where the traditional diet contains little red meat. Rheumatoid arthritis is also less common in vegetarians and vegans who, in turn, are less likely than meat-eaters to have an acid-forming diet. And it's less common in people who eat more cruciferous vegetables (such as cabbage and broccoli) and fruit:

A US study found rheumatoid arthritis was less likely in women who ate more fruit and cruciferous vegetables.

*(American Journal of Epidemiology, 2003)*

- **Action:**
**Eat an alkali-producing diet (see pages 84–85).**

People with rheumatoid arthritis are more likely to have antibodies to milk, cereal, eggs, fish, and pork. Food sensitivity is encouraged by an acid-producing diet. So it's possible that such a diet might encourage rheumatoid symptoms as part of an immune response triggered by a food sensitivity (see "Food allergy," pages 144–145). All this suggests that an acid-forming diet encourages rheumatoid arthritis.

Osteoarthritis can be linked with acidosis. The kidneys excrete acidic salts, but if this happens too slowly, the salts can crystallize in joint fluid and trigger inflammation:

> Research at the University of Padova in Italy shows that joint fluid in people with osteoarthritis may contain crystals, including those of calcium pyrophosphate dihydrate (present in one in three), apatite (present in one in four) and silicon dioxide.
>
> (Journal of Rheumatology, 2008)

> Irish researchers discovered that deposits in joints of crystals of octacalcium phosphate, tricalcium phosphate, carbonate-substituted hydroxyapatite, and magnesium whitlockite can be to blame for both osteoarthritis and the inflammatory joint condition *calcific periarthritis*.
>
> (Current Rheumatology Reports, 2003)

Gout causes arthritis and can be triggered by acidosis.

**DID YOU KNOW?**

Osteoarthritis pain worsens over the years as the "wear and tear" of everyday movement and weight-bearing roughens the smooth cartilage that covers the ends of the bones in each joint. Rheumatoid arthritis is an autoimmune disease, meaning it results from abnormal behavior by certain cells in the body's immunity system.

# Asthma

Food allergy triggered by an acid-producing diet can be responsible for asthma (see "Food allergy," pages 144–145). Pre-existing acidosis also makes airway-widening medication less effective early in an attack.

Airway tightening during an attack triggers deep rapid breathing. Too much carbon dioxide is then exhaled, leading to respiratory alkalosis. This triggers the bicarbonate buffering system to normalize the alkalosis. But if a person's diet is acid-producing, they could be short of bicarbonate, in which case their alkalosis might continue or even worsen. Breathing exercises can help:

> An Australian study found that shallow nose breathing ("Buteyko breathing"), or steady nose or mouth breathing (plus upper-body exercises, relaxation and good posture), during an attack decreased reliever-inhaler use by 86 percent, and halved the dose of inhaled-steroid medication.
> *(Thorax, 2006)*

Sodium bicarbonate can relax airways and encourage them to respond to airway-widening medication.

> Intravenous sodium bicarbonate improves blood pH and carbon-dioxide concentration in children with life-threatening asthma.
>
> *(Chest, 2005)*

• **Action:**
**Eat an alkali-producing diet (see pages 84–85).**

**Consider taking sodium bicarbonate.**

# Bad breath

An acid-forming diet encourages bad breath from residues of meat or refined carbohydrate on or between the teeth. Mouth bacteria can break these down and release unpleasant-smelling compounds.

The cells of people who eat a high-protein, low-carbohydrate diet, or have uncontrolled diabetes, must produce energy from fats instead of sugar. This leads to a build-up of acidic ketones in the blood. These taint breath with the scent of "pear drop" sweets or nail-polish remover.

A study at Duke University found that 63 percent of people on a high-protein, low-carbohydrate diet reported bad breath.

*(American Journal of Medicine, 2002)*

- **Action:**
Rinse your mouth with ½ teaspoon of sodium bicarbonate in a glass of water.

Eat a well-balanced, alkali-forming diet.

Seek urgent medical help if your breath smells of ketones and you have diabetes or suspect you might have it.

## DID YOU KNOW?

Over-the-counter breath fresheners such as mints, gum, breath sprays, and mouthwashes only mask the problem of bad breath, whereas sodium bicarbonate has the effect of neutralizing unpleasant odors.

# Cancer

Early research increasingly supports the idea that acidosis can encourage cancer. For example, studies indicate that cancer cells sometimes stop multiplying when the pH is relatively alkaline (just above 7.4). This pH also encourages more oxygen to enter cancer cells—which is good because low oxygen encourages cancer cells to multiply. Other research shows that cancer cells engender acidosis around them, which makes weak-alkali anti-cancer drugs less effective.

Several links suggest that acidosis encourages cancer:

- Obesity is associated with more bowel, gallbladder, kidney, and prostate cancers. Researchers also speculate that acidosis encourages obesity. So acidosis may prove to be a factor behind both obesity and cancer.

- A high protein intake encourages bowel and prostate cancer, whereas eating plenty of vegetables discourages bowel, breast, and stomach cancer. Similarly, an acid-producing diet encourages acidosis. So acidosis may prove to encourage certain cancers.

- Studies suggest that acidosis encourages diabetes. People with diabetes have an increased risk of certain

• Action:
Eat an alkali-producing diet (see pages 84–85).

cancers (for example, liver and pancreatic cancer). So acidosis may prove to encourage both diabetes and cancer. Researchers report that:

Giving sodium bicarbonate to mice with cancer alkalinizes the area around the cancer.

*(British Journal of Radiology, 2003)*

If the above also applies to humans, it would enable weak-alkali anti-cancer drugs to work. What's more:

Oral sodium bicarbonate increases cancer pH, reduces cancer growth, inhibits metastases (secondary cancers), and discourages lymph-node involvement in mice with cancer.

*(Cancer Research, 2009)*

Unproven treatment claims include:

- Applying sodium-bicarbonate paste to a rodent ulcer (basal-cell carcinoma) or squamous-cell skin cancer.

- Taking ascorbic acid and sodium bicarbonate for stomach or bowel cancer. These release sodium ascorbate and the hope is that this will damage cancer cells.

- Taking sodium bicarbonate mixed with maple syrup.

Finally, bicarbonate is proving useful in diagnosing cancer and monitoring treatment. This is because cancer cells convert bicarbonate to carbon dioxide, and magnetic resonance imaging (MRI) scans can monitor changing carbon-dioxide levels.

## Cataracts

These could be linked with acidosis as they can result from aging or stress encouraging a build-up in the eyes' lenses of insoluble calcium salts of phosphoric or uric acid. Taking sodium bicarbonate helps to dissolve these salts, as does a potassium-rich diet.

- Action:
  Eat an alkali-producing diet, as this is rich in potassium and its metabolism produces bicarbonate.

### Potassium power!

In these days of eating carry-outs and ready-prepared meals, many of us have far too little potassium in our diets. Potassium-rich foods are plentiful but are best eaten fresh, if appropriate, or else cooked as lightly as possible. The following are particularly good choices: white beans; adzuki beans; baked potatoes (with skins); steamed spinach, chard, and beet greens; baked squash; halibut; salmon; plain yogurt; dried apricots and dates; nuts; avocados; bananas.

# Colic in babies

Many parents use gripe water to soothe their baby's "colic" (restlessness, crying, and hiccups attributed to intestinal spasm). All manufactured brands contain sugar in some form plus small doses of ingredients such as sodium bicarbonate, fennel, dill, and ginger.

Gripe water is not approved as a medicine by the US Food and Drug Administration because imported brands may contain alcohol and sodium bicarbonate in amounts considered unsafe. Instead, manufacturers market it as a dietary supplement.

• **Action:**
If giving gripe water, check it contains no alcohol, follow the directions, and consult a pediatrition first if your baby is on any other medication.

# Confusion

This can be a symptom of severe metabolic acidosis—caused, for example, by serious heart, lung, or liver disease.

- **Action:**
  **Seek urgent medical help.**

# Convulsions

Convulsions occur in 70–80 percent of people with epilepsy, despite their medication, and acidosis may be a factor. Most people with epilepsy go into remission but even with optimal medication, 20–30 percent have repeated convulsions, and half have more than one a month. The causes are unclear, but among the suggestions are nutritional factors such as low-grade chronic metabolic acidosis, and various vitamin and mineral deficiencies.

- **Action:**
  **Eat an alkali-producing diet (see pages 84–85).**

> Researchers believe the chronic low-grade metabolic acidosis associated with most modern Western diets encourages chronic epilepsy by over-exciting brain cells.
> *(Epilepsy and Behavior, 2006)*

Convulsions can also result from a slow-onset immune reaction caused by a food sensitivity enabled by acidosis (see "Food allergy," pages 144–145).

# Cystitis

A person with acidosis produces relatively acidic urine. This can inflame the bladder and encourage urine infections. The resulting cystitis causes painful and frequent urination. The urge to urinate may be almost continuous and the pain itself is often sharp and burning.

Alkalinizing the body with sodium bicarbonate helps to neutralize excess acid in the urine. It also kills bacteria and makes certain antibiotics more effective.

If left untreated, cystitis may lead to a kidney infection so it's important to start treatment at the first signs.

Men and children who develop the symptoms of cystitis should always see a doctor. Women should see a doctor if it's the first time they have had cystitis, or if they have had it three or more times in a year.

• **Action:**
Affected women should consider drinking ½ teaspoon of sodium bicarbonate in a glass of water four times a day. Other alkalinizing remedies are available from drugstores.

See a doctor if this is your first attack, or you are no better within 2–3 days.

# Depression

Depression is sometimes reported as a side-effect of a high-protein, low-carbohydrate diet, which in turn is associated with low-grade metabolic acidosis.

Depression can also result from a slow-onset immune reaction caused by a food sensitivity enabled by acidosis (see "Food allergy," pages 144–145).

• **Action:**
Eat an alkali-producing diet (see pages 84–85).

# Diabetes, pre-diabetes, and metabolic syndrome

Researchers believe that acidosis encourages cells to become resistant in insulin. Normally, this hormone enables blood sugar to enter cells. But if cells are resistant, blood-sugar rises. The pancreas then produces extra insulin to prevent high blood sugar. This "pre-diabetes" can have adverse effects, including obesity, fluid retention, high blood pressure, raised LDL-cholesterol, fatigue, faintness, mood swings, dry skin, skin tags, and darkened skin.

Some people with pre-diabetes have a collection of problems together called the metabolic syndrome. This greatly encourages diabetes, heart disease, and strokes and is characterized by having three of the following: insulin resistance, excess fat around the waist, high blood pressure, high blood fats, or low HDL-cholesterol.

A study at the University of Texas reveals that overly acidic urine (which indicates acidosis) is a feature of the metabolic syndrome and associated with the degree of insulin resistance.
(Clinical Journal of the American Society of Nephrology, 2007)

Continued insulin resistance can exhaust pancreatic cells, resulting in type 2 diabetes. People with diabetes tend to have overly acidic urine, indicating acidosis.

• **Action:**
In addition to whatever other treatment you need, eat an alkali-producing diet (see pages 84–85)

Researchers at the University of Texas say that excess weight and an acid-producing diet can't entirely account for the overly acidic urine in people with diabetes.
*(Journal of the American Society of Nephrology,* 2006)

Dieters may develop acidosis if their diet contains insufficient vegetables and fruit. This is a particular problem because many people with pre-diabetes, metabolic syndrome, or diabetes are overweight and repeatedly trying to slim.

Before oral anti-diabetic drugs, and insulin, doctors often treated diabetes with sodium bicarbonate.

## *Diet with alkali-producing foods*

*If you want to lose weight, be aware that for the same calorie intake you can eat a far larger amount of vegetables and fruit than of foods rich in fat, protein, and refined carbohydrates. For example, you could swap one 9 oz (250 g) fat- and sugar-free muffin for 2 lb (900 g) pineapple, half a melon, 2 pears, 5 oz (150 g) grapes, half a kiwi fruit, half a papaya, and 2 wholemeal rolls! A snack of 200 g of cashew nuts would be the equivalent of 8 baked jacket potatoes (without butter or fillings, of course).*

# Diarrhea

If persistent severe diarrhea originates in the small intestine, it can cause metabolic acidosis from the loss of bicarbonate. If it originates in the large intestine it can cause metabolic alkalosis from the loss of chloride.

Diarrhea can also result from a slow-onset immune reaction enabled by acidosis (see "Food allergy," pages 144–145).

• Action:
Eat an alkali-producing diet (see pages 84–85).

Consider taking oral rehydration salts (from a pharmacy) or ½ teaspoon of sodium bicarbonate in a glass of water up to four times a day.

# Dizziness, light-headedness and fainting

These can be associated with either respiratory or metabolic alkalosis (see pages 103–105).

They can also result from the acidosis that can accompany pre-diabetes and a slimming diet (particularly one low in carbohydrate).

• Action:
Check out these and other possible causes and treat as appropriate. Also, eat an alkali-producing diet (see pages 84–85).

# Drowsiness

This could be a sign of respiratory or metabolic acidosis.

• Action:
For causes and treatments, see pages 100–103.

# Fatigue and poor concentration

These can result from metabolic acidosis. One possible reason is exhaustion from rapid breathing triggered as the body tries to correct a low pH.

A second possible cause is going on a slimming diet, particularly if it is one low in slowly absorbed carbohydrate. This can lead to a shortage of sugar for energy production and mean that cells must use protein or fat as an energy source instead. But converting these nutrients to usable fuels takes longer than converting carbohydrate to sugar. This is a particular disadvantage for brain cells, as it can result in fatigue and poor concentration. In addition, burning fat for energy produces acidic ketones. Not only can the resulting acidosis cause fatigue but it also encourages pre-diabetes, which can trigger tiredness. The lack of sugar can also affect heart muscle (see "Heart disease," pages 149–151), causing low energy.

A third possible reason is a slow-onset immune reaction associated with a food sensitivity enabled by acidosis (see "Food allergy," pages 144–145).

• Action:
Eat an alkali-producing diet (see pages 84–85).

## DID YOU KNOW?

The whole grains, peas, and beans in an alkaline diet release their sugars slowly. This helps prevent mental and physical fatigue by providing the brain and muscles with a steady supply of fuel.

# Fibromyalgia

With this painful and chronic condition you have stiff, weak, knotted shoulder and back muscles, and tender points on hips, knees, neck, spine, elbows, or buttocks. The pain is heightened when pressure is applied to these areas. Further symptoms may include fatigue, poor sleep, headaches, dizziness, numbness, tingling, irritable bowel and bladder, restless legs, poor memory and concentration, depression, anxiety, over-sensitivity to noise, light, and temperature, and Raynaud's phenomenon (when the blood flow to the fingers or toes is restricted, resulting in a loss or change in color).

Although the causes of fibromyalgia remain medically unexplained, many experts blame metabolic acidosis, suggesting that this deposits acidic ions in muscles and connective tissue. Various studies support this idea, including:

A study at the University of Oslo in which most of the volunteers with fibromyalgia who ate a vegetarian diet for 3 weeks reported less pain and better well-being.

*(Plant Foods for Human Nutrition, 1993)*

- **Action:**
Eat an alkali-producing diet (see pages 84–85).

# Food allergy

Chronic low-grade metabolic acidosis is associated with low bicarbonate in the body. This means pancreatic juice may not contain enough bicarbonate to alkalinize acidic food residues entering the duodenum. As a result, its enzymes may not break down proteins properly; also, acidic food residues can damage the intestinal lining. Whole undigested protein molecules can then pass through the "leaky" lining into the blood and sensitize the immune system.

When someone who has become sensitized to a protein next eats it, one of two things can happen. The first is that immunoglobulin E (IgE) antibodies can adhere to it in the intestine, blood, or elsewhere, forming tiny particles called immune complexes, which can cause trouble (for example, by blocking tiny blood vessels). The second is that white cells can release inflammatory substances (such as histamine and leucotrienes), which can increase the production of potentially damaging free radicals (reactive oxygen-containing ions).

Food-allergic symptoms beginning within minutes or up to 2 hours are caused by a fast-action immune response. This can result from eating even a little of the particular food. Common culprits are milk, eggs, peanuts, tree nuts, fish, shellfish, soy, wheat, and oranges. Possible symptoms include swollen face, lips, mouth, tongue, and throat; vomiting; diarrhea; abdominal pain;

• Action:
Eat an alkali-forming diet.

Consider taking sodium bicarbonate during an attack (though not if there are symptoms of fluid retention).

itching; allergic rhinitis ("hayfever"); urticaria (hives or "nettle-rash"); asthma; and conjunctivitis. Pre-existing eczema may worsen. At worst there is anaphylactic shock (with potentially fatal breathing difficulty, a fall in blood pressure and, perhaps, loss of consciousness).

Food-allergic symptoms beginning later, though within 72 hours, are caused by a slow-onset immune response. Affected people may react to several foods. Possible symptoms include flushing, nausea, vomiting, diarrhea, esophagitis (inflamed gullet), gastritis (inflamed stomach), abdominal pain, fatigue, muscle weakness, aching and stiffness, eczema, joint pains, palpitations, fluid retention (perhaps with bloating, headaches, fluctuating weight, temporarily raised blood pressure, depression, convulsions, and restless legs), and gallstones.

## Food intolerance

*Unlike food allergies, food intolerances don't provoke antibody production. Certain types, such as lactose (milk sugar) intolerance, occur because the enzyme needed to digest a nutrient is lacking. Others include genetic susceptibility to various foodstuffs (such as to monosodium glutamate, causing the headache, flushing, and other symptoms of "Chinese restaurant syndrome").*

## *Fungal skin and nail infections (including Athlete's Foot):*

*Baking soda is known to be an effective antifungal treatment. As it also softens and soothes the skin, it can be particularly effective in treating Athlete's Foot. Try soaking feet in a bowl of water containing a handful of sodium bicarbonate for 20 minutes daily.*

*Alternatively, rub a paste of sodium bicarbonate and water into affected toenails or fingernails and leave on for 20 minutes. Also, sprinkle sodium bicarbonate between the toes each day.*

# Gallbladder disease

The risk of gallstones or an inflamed gallbladder rises in people who lose weight rapidly, as this can cause acidosis. The risk of gallstones also rises in people who are obese, and certain experts believe that acidosis encourages obesity. None of this proves that acidosis causes gallstones, but it does suggest that there may be a possible link .

On a different tack, acidosis encourages an immune response to certain proteins. People with gallstones have a raised risk of food sensitivity (see "Food allergy," pages 144–145), the most likely culprits being eggs, pork, onions,

• Action:
Eat an alkali-producing diet (see pages 84–85).

chicken, milk, and coffee. So acidosis might indirectly encourage gallstones. One suggested mechanism is that a food-sensitive immune response inflames the bile duct, making bile stagnate in the gallbladder, and thus encouraging the formation of gallstones.

# Gout

Gout is associated with needle-like crystals of sodium urate (specifically, monosodium urate monohydrate) in joints or under the skin. Urates form when purines from food and from the body's cells are broken down by the body into uric acid, which is carried in the blood as urate. The kidneys normally excrete excess urate, but with acidosis they may not do so fast enough, so urate levels rise.

Some people report that taking sodium bicarbonate alleviates or cures an attack of gout. By reducing acidosis, this presumably (it is not yet scientifically proven) enables more sodium urate to dissolve in the blood, encouraging urate crystals in the joints to dissolve. Gout attacks are more likely to occur at night, which is when the body is particularly acidic.

In an online poll of people with gout, 85 percent said that taking sodium bicarbonate helped, sometimes within 24 hours of first taking it and often within a week.

- **Action:**
Consider taking ½ teaspoon of sodium bicarbonate four times a day for a week.

Eat an alkali-producing diet (see pages 84–85).

# Headaches

Headaches are a possible symptom of metabolic acidosis with a pH below 7.35. It's also possible, though unproven, that chronic low-grade metabolic acidosis can cause headaches as a result of the body's buffer systems working extra hard to keep the blood's acid–alkali balance within its normal tightly controlled range.

Another cause of headaches is a slow-onset immune reaction associated with a food sensitivity enabled by acidosis (see "Food allergy," pages 144–145).

If you suffer from repeated headaches, it's worth examining your diet. For example, the US Dietary Guidelines recommend eating from 5 to 13 servings of vegetables and fruit each day, but it is known that more than half the adult population in America fail to do this.

• **Action:**

If you eat an acid-producing diet or have other reason to believe you might have acidosis, consider taking ½–1 teaspoon of sodium bicarbonate in a glass of water to help cure a headache.

Eat an alkali-producing diet (see pages 84–85).

## Maintain your fluid levels

*Prevent dehydration headaches by drinking enough water each day to keep your urine very pale yellow. The actual volume you need varies with your state of health, age, activity level, and with the heat and humidity of your environment.*

# Heart disease

The findings below indicate that acidosis encourages low energy, an irregular heartbeat (atrial fibrillation), chest pain on exercise (angina), heart attacks, and heart failure. Whether chronic low-grade metabolic acidosis (as with a typical Western diet) has similar effects isn't yet known.

Many studies show that acidosis can:

- Reduce energy production from sugar in heart-muscle cells; energy production from fats and proteins takes longer than from sugar, so low energy is an early problem.

- Irritate the coronary arteries (if the pH is below 7.35), encouraging tiny tears. LDL-cholesterol then seeps into the artery lining and attracts white blood cells. These white cells provoke inflammation, which oxidizes cholesterol. They then engulf the oxidized cholesterol. Calcium infiltrates the damaged artery walls, while smooth-muscle cells produce collagen to cover the leaks. All this causes atherosclerosis—stiffening of the arteries plus narrowing by a chalky, fatty, fibrous, white-cell-laden substance called atheroma building up in the artery walls. Patches of atheroma can rupture, encouraging a blood clot, which could then trigger a heart attack.

• Action:

**Eat an alkali-producing diet (see pages 84–85).**

- Interfere with the passage of potassium, sodium, and calcium across cell membranes, and also rob the body of potassium, calcium, and magnesium as they neutralize acids in the urine. This weakens heart-muscle, prevents the normal conduction of electrical messages, and makes sodium and calcium accumulate in blood, encouraging high blood pressure and the development of atheroma along the artery walls.

- Increase LDL-cholesterol (the sort that's potentially dangerous if oxidized) and decrease HDL-cholesterol (the protective sort).

- Increase adrenaline, which speeds the heart and boosts its cells' need for oxygen.

- Encourage insulin-resistance; blood sugar then can't enter cells normally, so rises, triggering extra insulin production. But high insulin has adverse effects, such as boosting LDL-cholesterol.

- Raise fibrinogen, which can encourage blood clots, which can block a coronary artery.

> **DID YOU KNOW?**
>
> Eating an alkaline diet is a simple and easy way of preventing or reducing any inflammation in your coronary arteries, and thereby helping to ward off a heart attack. So each time you eat a serving of vegetables or fruits, for example, you are doing your heart a good turn.

For example:

> Acidosis rapidly reduces the passage of calcium into
> heart-muscle cells. This prevents the heart pumping
> efficiently and rhythmically. The greater the acidosis, the
> less well the heart pumps.
>
> *(Ciba Foundation Symposium,* 1982*)*

Blood pH is lowest (at its most acidotic) during sleep,
which is when fatal heart attacks are most common.

Palpitation can result from a slow-onset immune
reaction associated with a food sensitivity enabled by
acidosis (see "Food allergy," pages 144–145).

## *A healthy heart diet*

*An easy way to improve your diet and thereby discourage heart disease is to
eat more vegetables. These make your body more alkaline, supply natural anti-
inflammatories, and reduce your appetite for acid-producing foods. Cucumber
and sprouted seeds are among the top alkali-producers, but all vegetables are
helpful (see pages 84–85).*

# High blood pressure

Acidosis seems to encourage high blood pressure:

• **Action:**
**Eat an alkali-producing diet (see pages 84–85).**

In a study of 15,385 women, those whose diets had a higher potential acid load had higher blood pressure than those with a lower load.

*(Hypertension,* 2009)

The probable culprit is a lack of vegetables and fruit, since there is no evidence that a high protein intake in itself raises blood pressure.

Taking sodium bicarbonate (or any other alkalizer) may reduce blood pressure:

Consuming sodium bicarbonate as part of a low-salt diet can lower blood pressure by reducing calcium excretion in urine.

*(Journal of Hypertension,* January 1996, Vol. 14, Issue 1*)*

Acidosis makes cells more resistant to insulin. The pancreas then produces extra insulin to prevent high blood sugar. This makes arteries over-sensitive to adrenaline, thereby encouraging high blood pressure.

If you have high blood pressure and are also overweight, losing weight could be a highly successful treatment. In a review of studies involving 4,874 people, blood pressure fell by 1 point for each 2 pounds lost.

# Indigestion, heartburn, gastritis, and peptic ulcer

Acidosis from an acid-producing diet increases the stomach's production of bicarbonate *and* acid. The acid encourages "acid" indigestion, heartburn, peptic ulcers and gastritis (inflamed stomach lining). When the stomach contents enter the duodenum, pancreatic juice provides bicarbonate to neutralize the acid. This stimulates stomach-lining cells to produce more bicarbonate; at the same time they automatically produce even more acid.

Taking sodium bicarbonate can neutralize excess acid. But its sodium is absorbed into the blood and some people are believed to be sodium-sensitive. So antacids such as calcium carbonate or magnesium or aluminum hydroxide are probably preferable (and scarcely absorbed). Antacid use has declined anyway with the advent of acid-suppressant medication.

Excess acid in the digestive system is also said to encourage an inflamed colon (colitis).

Note that too little stomach acid can also cause indigestion. While some people with gastritis or a peptic ulcer make too much acid, most don't, and some make too little. Indeed, low acid is a major cause of peptic ulcers and gastritis. This is because it encourages *Helicobacter pylori* bacteria to inflame the stomach-lining cells and hamper their production of protective bicarbonate-containing

• **Action:**
Consider taking ½–1½ teaspoons of sodium bicarbonate in a glass of water. Alternatively, take another antacid or discuss with your doctor or pharmacist whether to take an acid-suppressant.

mucus. It also encourages stomach cancer. Around two in five of us are infected, though only one in ten infected people develop an ulcer.

> Japanese research strongly correlates low stomach acid with increased rates of *H. pylori* infection.
> *(Biotechnic and Histochemistry, 2001)*

Lastly, a slow-onset immune response associated with a food sensitvity enabled by acidosis (see "Food allergy," pages 144–145) can cause gastritis and esophagitis (inflamed gullet).

## How to test your stomach acid

*To help identify whether you suffer from poor stomach-acid production, drink a teaspoon of sodium bicarbonate in a glass of water on an empty stomach. If you don't belch within 5–10 minutes (from acid converting sodium bicarbonate into carbon dioxide) you may have low acid, in which case antacids are unlikely to relieve indigestion.*

# Infections

It is suggested that increasing the blood's alkalinity helps to prevent or treat bacterial and viral infections. Although there may be anecdotal evidence of this, it has yet to be scientifically proven.

• **Action:**

If you suspect that you may have an infection, drink ½ teaspoon of sodium bicarbonate in a glass of water four times a day.

Eat an alkali-producing diet (see pages 84–85).

# Infertility

To be successful, a sperm must first undergo a process called capacitation—the alteration of the surface of its head so it can adhere to an egg. This is done by fertilization-promoting peptide, a substance produced by the prostate and mixed with sperm on ejaculation.

Capacitation occurs in the cervix or womb but occurs only if the alkalinity is right. This happens just before ovulation, when cervical glands produce "sperm-friendly" mucus that is suitably alkaline, abundant, clear, elastic, and watery, and contains strands that are aligned to encourage sperm penetration.

• **Action:**
Eat an alkali-producing diet to encourage favorable alkalinity.

# Inflammation

Inflammation is part of many disorders, and research suggests it is encouraged by acidosis. For example:

- **Action:**
**Eat an alkali-producing diet (see pages 84–85).**

One study found a high-protein, low-carbohydrate diet (which is known to be associated with acidosis) increases C-reactive protein—a reliable marker of inflammation.

(*Angiology*, 2000)

# Kidney disease

Acidosis can both result from kidney disease and encourage it.

- **Action:**
**Eat an alkali-producing diet (see pages 84–85).**

A study of 1624 women in the US Nurses' Health Study found that a higher intake of protein—particularly non-dairy animal protein—encouraged greater decline in kidneys already slightly impaired.

(*Annals of Internal Medicine*, 2003)

A review from the Indiana University School of Medicine associates a high-protein intake with increased urine protein, sodium, and potassium, and faster progression of chronic kidney disease. Such disease is often symptom-free, so anyone contemplating a high-protein diet should first have a blood test for creatinine and a urine test for protein. There are no clear contra-indications if this shows that the kidneys are healthy.

(*American Journal of Kidney Disease*, 2004)

**Alkalinizing the diet may slow the progression of chronic kidney disease:**

A study at the Royal London Hospital found that kidney disease progressed rapidly in only 9 percent of people treated with sodium bicarbonate, compared with 45 percent of others. Those who took sodium bicarbonate were also less likely to need dialysis.

(*Journal of the American Society of Nephrology*, July 2009)

# Kidney stones

Calcium oxalate stones (the commonest type) and uric-acid stones (for example, in people with gout) are more likely with acidosis.

An Italian study of 187 people with calcium-containing stones found their diet's acid-producing potential was the biggest risk factor. The researchers suggest such people should eat plenty of vegetables and fruit but relatively little animal protein.

(*Urology Research*, 2006)

• **Action:**
Eat an alkali-producing diet (see pages 84–85).

Consider taking ½ teaspoon of sodium bicarbonate in a glass of water four times a day for a week, followed by ¼ teaspoon four times a day for 2 weeks.

Studies show stones are more likely with:

- Type 2 diabetes.

- Metabolic syndrome (see page 139).

- Obesity.

Each can be associated with acidosis.

Consuming table salt—which is acid-producing—encourages the formation of calcium oxalate and calcium phosphate stones.

Sodium bicarbonate can alkalinize urine so it is more able to rid the body of uric acid without forming stones. Alternative urine alkalinizers include potassium citrate.

## DID YOU KNOW?

Kidneys stones are four times as likely in men as in women. Many are flushed from the kidneys down into the bladder and out of the body without causing symptoms. But the larger a stone, the more likely it is to cause pains called renal colic, and to get stuck.

# Mouth ulcers

Sodium bicarbonate can soothe these painful aphthous ulcers and aid their healing. More commonly known as "canker sores," these are not the same as "fever blisters," which are caused by the *herpes simplex* virus and require different treatment.

- **Action:**
  Dissolve 1 teaspoon of sodium bicarbonate in a glass of water and swirl a mouthful around your mouth every 2–3 hours.

# Muscle cramp

Acidosis can draw magnesium from the body to
accompany acidic anions, such as lactate, in the urine,
encouraging cramp.

Simple measures for preventing cramp include taking
moderate daily exercise, keeping warm, and eating a
healthy diet with neither too much nor too little salt. But
if cramp is frequent or occurs when walking or with a
repeated movement, it's wise to ask your doctor to check
out all possible causes

- **Action:**
  Rest as necessary to
  allow your blood to
  clear its acidic lactate.

  Eat an alkali-producing
  diet.

  Consider taking ½
  teaspoon of sodium
  bicarbonate every
  6 hours if you get
  repeated cramp.

# Muscle fatigue and aching

Weakness and rapid fatigue during intense exercise are
traditionally explained as being caused by muscle cells
producing acidic lactate ions. Certainly during intense
exercise the pH of muscle cells can fall from 7 to 6.8. But
research shows that what actually happens is:

During intense exercise, muscle cells eventually have to
produce energy without using oxygen, which releases
hydrogen ions and lactate. Hydrogen ions encourage
cellular and generalized acidosis. Lactate can also move
into the blood.

*(American Journal of Physiology – Regulatory,
Integrative and Comparative Physiology, 2004)*

- **Action:**
  Eat an alkali-producing
  diet (see pages 84–85).

  Consider taking
  sodium bicarbonate,
  but only with
  supervision by a
  qualified coach or a
  doctor.

If muscles didn't produce lactate, acidosis, and fatigue would occur faster and performance would be severely impaired. This is because lactate is a very useful fuel for muscles as it produces energy so fast. Indeed, some athletes take lactate in a fluid-replacement drink before, during or after exercise. During exercise, the muscle pain known as "the burn" is associated with a build-up of hydrogen ions and signals that lactate is enabling the rapid production of energy.

Most exercise-associated muscle cramp results from acidosis over-exciting nerve receptors in muscles. "Delayed-onset" aching the next day results from muscle damage and exercise-induced inflammation.

Taking sodium bicarbonate before intense exercise can buffer acidosis, enabling more prolonged exercise. This "soda loading" can reduce muscle fatigue and enhance recovery. But the large amounts often recommended (about 0.3g per kg body-weight) can cause symptoms (see pages 124–125) and shave only seconds off performance time. So it is controversial.

A research review noted that performance improved for 400–800m runners, 200m swimmers, cyclists, and boxers who took sodium bicarbonate before an event.
(*British Journal of Sports Medicine*, 2010)

### DID YOU KNOW?

You can make your own sports drink to replenish the water, sodium, and carbohydrate your body uses as you exercise. Simply stir a small pinch of salt into a pint of fruit squash, or into a mixture of half a pint of water and half a pint of fruit juice.

Muscle fatigue and aching unrelated to exercise can result from a slow-onset immune reaction associated with a food sensitivity enabled by acidosis (see "Food allergy," on pages 144–145).

# Muscle wasting

Aging increases acidosis and is also associated with muscle wasting. The chronic low-grade acidosis that is so common in Western societies is thought to encourage muscle wasting.

• **Action:**
**Eat an alkali-producing diet (see pages 84–85).**

> Research at Tufts University, Boston, involving 384 over-65s, suggested a higher intake of foods rich in potassium, such as vegetables and fruit, helps prevent age-related muscle wasting.
>
> *(American Journal of Clinical Nutrition,* 2008)

If an elderly person's limbs are looking increasingly frail and thin, it's very likely that their muscles are gradually wasting away through lack of exercise. Encourage them to take a gentle walk each day and seek out activities that involve stretching and movement.

# Nausea and vomiting

This can be caused by metabolic acidosis (see pages 100–102) with a pH below 7.35.

It is also possible, though unproven, that chronic low-grade metabolic acidosis (see page 106) can cause these symptoms too.

Nausea and vomiting can also be symptoms of alkalosis (see pages 103–105).

Another cause is a slow-onset immune reaction associated with a food sensitivity enabled by acidosis (see "Food allergy," pages 144–145).

- **Action:**
  If you suspect acidosis, consider taking ½ teaspoon of sodium bicarbonate in a glass of water every 3 hours.

  Eat an alkali-producing diet (see pages 84–85).

  If symptoms persist for more than a few days, consult a doctor.

## *Hangover cure*

*Too much alcohol causes nausea, headache, and other debilitating symptoms in all but 25–30 percent of imbibers. It's commonly believed that dehydration is the problem, and too much alcohol does indeed increase urination, with just 9 fl oz (250 ml) of alcohol triggering the passing of more than 2 pints (1 liter) of urine. It also reduces the body's blood sugar and bicarbonate levels so drinking ½ a teaspoon of sodium bicarbonate dissolved in a glass of water may help to speed up recovery.*

# Numbness and tingling

These can be symptoms of either respiratory or metabolic alkalosis (see pages 103–104).

Poor circulation in the peripheral blood vessels is a common cause of numbness in the fingers, toes, nose, and ears when a person feels cold. Chronic low-grade metabolic acidosis is nowadays thought to encourage poor circulation, which can result in such susceptibility. So if you have this kind of numbness, you should consider your diet (see pages 80–87).

- Action:
Consult your doctor if the symptoms continue and you don't know the cause.

## *How to ease hyperventilation*

*If you have symptoms that are associated with rapid breathing (hyperventilation) triggered by anxiety or pain, consider seeing whether it helps to breathe in and out of a large paper bag to raise your blood's carbon-dioxide level. You can then try to keep the symptoms at bay by breathing more slowly.*

# Osteoporosis

Osteoporotic bone is light and fracture-prone. The risk factors include age; too much or too little exercise; smoking; too little bright outdoor light; a lack of dietary calcium, magnesium, zinc, vitamins C, D, and K, plant hormones, or fiber; too much "diet cola" (because of its phosphoric acid content); early menopause; anorexia; and various medications and illnesses (including gut and thyroid disorders—some of which can result from acidosis).

Studies also implicate inflammation, which can result from acidosis.

Acidosis also impairs bone health by decreasing the activity of bone-building cells (osteoblasts) but increasing the activity of bone-destroying cells (osteoclasts).

Research suggests that acidosis may draw calcium from the bones to partner acidic anions in the urine.

Finally, researchers believe that chronic low-grade metabolic acidosis from an acid-producing diet can be a factor. In particular, they point to a diet lacking in vegetables and fruit:

> This review of acidosis strongly suggests that diet-induced chronic low-grade metabolic acidosis has significant adverse effects that might be counterbalanced by an alkali-producing diet.
>
> *(British Journal of Nutrition, 2010)*

- **Action:**
Eat an alkali-producing diet, in particular plenty of vegetables and fruit.

A review from the University of Illinois suggests the acid-producing effects of dietary protein are minor compared with the alkalinizing effects of vegetables and fruits.

*(American Journal of Clinical Nutrition, 2008)*

Vegetables and fruit contain many "bone-friendly" nutrients and have an alkali-producing effect.

Several studies suggest an alkaline supplement could have a similarly helpful alkalinizing effect. For example:

A study at Tufts University, Boston, of 171 people aged 50-plus, reports that taking a supplement of sodium or potassium bicarbonate for 3 months reduced calcium and bone-turnover markers in urine.

*(Journal of Clinical Endocrinology and Metabolism, 2009)*

An alkali-producing diet helps to prevent and treat osteoporosis. Each of us needs between 5 and 13 servings of vegetables and fruit a day, depending on our weight and activity level. One serving is a medium fruit, a cup of green leafy vegetable, or a half cup of other vegetable.

## DID YOU KNOW?

Bright outdoor daylight protects against osteoporosis in two ways. Its ultraviolet rays enable vitamin-D production in the skin. And its brightness triggers the retinas to stimulate the hypothalamus, so increasing estrogen. Both vitamin D and estrogen strengthen bones.

# Overweight and obesity

Acidosis encourages resistance to the hormone insulin.
Normally, insulin enables blood sugar to enter cells. But
if cells are resistant, blood sugar rises and the pancreas
produces extra insulin to convert surplus sugar into fat.

Many conditions are encouraged by, or associated
with, both obesity and acidosis. For example, obesity
and acidosis each encourage heart disease, high blood
pressure, strokes, diabetes, metabolic syndrome,
heartburn, gallstones, pancreatitis, osteoarthritis, gout,
kidney stones, asthma, fatigue, depression, sleep apnea
("stop-breathing" attacks during sleep), underactive
thyroid, absent periods, infertility, and certain cancers
(though any influence from acidosis is only suggested and
not scientifically proven).

Since acidosis also encourages obesity, it is tempting to
speculate that this is a linking factor.

A hypothesis from the University of Bochum in Germany
explains how obesity is related to acidosis and the
production of potentially damaging free radicals (reactive
oxygen-containing ions). It also explains how the latter
"oxidative stress" could be the link between obesity and
diseases commonly associated with it—such as high
blood pressure, diabetes, heart disease, and strokes.

*(Medical Hypotheses,* 2010)

• **Action:**
Eat an alkali-producing
diet (see pages 84–85).

**DID YOU KNOW?**

Eating plenty of green,
red, orange, yellow,
and purple vegetables
as part of a healthy diet
is an excellent way of
filling yourself up and
reducing your appetite
for foods laden with
calories and fat.
Another advantage is
that alkalinizing your
diet is likely to boost
your energy.

It would also be worth investigating whether acidosis is responsible for other conditions encouraged by obesity—including age-related macular degeneration (an eye disease), Alzheimer's disease, deep vein thrombosis, pulmonary embolism, pre-eclampsia (a pregnancy condition), high blood fats, polycystic ovary syndrome, and liver disease.

People who don't eat meat—and are therefore less likely to have chronic low-grade metabolic acidosis—are also less likely to be obese:

> A study of 37,875 people concluded that meat-eaters are more likely to be obese.
>
> *(International Journal of Obesity, 2003)*

Another fact is that most obese people are malnourished. Severe malnutrition decreases the bicarbonate content of pancreatic juice, encouraging acidosis.

Avoid a high-protein, low-carbohydrate diet, as this encourages acidosis. An added bonus of an alkali-producing diet is that you can eat more yet still lose weight, because eating more vegetables and fruit can turn a fat-storing tendency into a fat-burning one.

# Pain

Damaging factors such as inflammation, injury, heat, cold, and lack of oxygen can trigger pain receptors by releasing a variety of agents. These include hydrogen ions (protons) and adenosine triphosphate (from damaged cells), serotonin and prostaglandins (from mast cells), and cytokines and nerve growth factor (from white blood cells called macrophages). Hydrogen ions cause local acidosis. These various agents act on specific receptors and on ion channels in the endings of sensory receptors ("nociceptors"). These receptors then send nerve signals to the spinal cord and brain, eliciting pain.

Pain is felt when damaged tissue has a pH of 7 or below (depending on its site).

It's likely that "whole-body" (as opposed to local) acidosis encourages pain, too. So anyone troubled with pain of any sort—including headaches, period pain, sciatica, and tendinitis—might benefit from changing the way they eat.

• Action:
Eat an alkali-producing diet (see pages 84–85).

## *Eating to reduce pain*

*It may be tempting to try to soothe discomfort with the sweetness of cookies or candy. But it might be better to reach instead for some fruit or a creatively prepared salad. This is because the pain just might lessen if you make your body a tiny bit more alkaline.*

# Poor circulation

Research strongly suggests that acidosis encourages
the narrowing and stiffening of arteries known as
atherosclerosis (see "Heart disease," pages 149–151).
Although the body compensates for the obstruction to
normal circulation by developing high blood pressure,
atherosclerosis is eventually likely to reduce the rate of the
blood supply. This, in turn, can strain the normal working
of all the body's cells. Some of the most obvious of the
many possible symptoms are:

- Abnormal sensitivity to cold

- Impotence in men

- Lack of libido

- Paleness

- Poor memory

- Slow wound healing (especially on legs, ankles,
  and feet).

• **Action:**
**Eat an alkali-producing
diet (see pages 84–85).**

# Premature aging

Scientists have long searched for lifestyle factors that encourage a long life and discourage age-related conditions such as wrinkles, age spots, arthritis, heart disease, diabetes, cancer, osteoporosis, age-related macular degeneration, cataracts, poor memory, and Alzheimer's disease.

Long-lived peoples include certain groups in Russia (the Georgians), Pakistan (the Hunzas), Ecuador, China, Tibet, and Peru. One link is that they tend to live at high altitudes and drink mountain water rich in alkaline minerals such as calcium, which help prevent acidosis. Another is their consumption of fermented vegetables, fruit, milk, cereal grains, meat, or fish. These contain acids (such as lactic, acetic, and malic) produced during fermentation but metabolized to alkali in the body.

Another related fact is that spa water from medicinal springs is invariably alkaline.

In contrast, chronic low-grade metabolic acidosis is very common in peoples eating a typical Western diet. And aging exaggerates acidosis because declining kidney function reduces the excretion of excess acid:

- **Action:**
  Eat an alkali-producing diet (see pages 84–85).

## DID YOU KNOW?

We each have a better chance of aging well if we nourish ourselves carefully. Three tips are to choose wholegrain foods rather than those made with refined grains; eat vegetables and fruits of many hues; and have green leafy vegetables such as Brussels sprouts, cabbage, collard greens, kale, lettuce, and spinach each day.

A research review found that aging is accompanied by
worsening low-grade metabolic acidosis, with an increase
in blood pH and a decrease in its bicarbonate that
may reflect the expected age-related decline of kidney
function. Also, the blood's carbon dioxide concentration
falls with age, because the lungs compensate for acidosis
with more rapid breathing.

*(The Journals of Gerontology, 1996)*

The pH change accompanying acidosis has damaging
effects on cells and metabolism in general (see pages
94–97), some of which encourage premature aging.
For example, it is associated with structural changes in
proteins, the oxidation of fats, and a lower oxygen supply
to cells.

# Restless legs

This may result from a slow-onset immune reaction
associated with a food sensitivity enabled by acidosis
(see "Food allergy," pages 144–145).

• Action:
Eat an alkali-producing
diet (see pages 84–85).

Consider taking a dose
of sodium bicarbonate
(see page 122).

# Skin problems

Theoretically, at least, acidosis can trigger or worsen various skin conditions. This is because it encourages:

- Itching

- Inflammation, and/or pain from broken or inflamed skin

- Allergy

- Infection

- Raised insulin (see "Diabetes," pages 139–140), which encourages dry skin, and skin tags and dark skin on the neck, under the breasts, and in the armpits and groin

One way of treating many skin conditions is to eat an alkali-producing diet to prevent acidosis.

Applying sodium bicarbonate directly to affected skin combats certain problems. For example:

When 31 people with mild-to-moderate psoriasis took bicarbonate baths, almost all reported an improvement. The baths reduced itchiness and irritation so well that the volunteers continued their baths after the study.

*(Journal of Dermatological Treatment*, 2005)

• Action:
ALLERGIC CONTACT DERMATITIS: reduce itching by applying a paste of sodium bicarbonate and water (see page 126).

ECZEMA: have an alkaline bath (see page 126).

ITCHING: apply a paste of sodium bicarbonate and water, or have an alkaline bath (see page 126).

# First aid

- CUTS AND GRAZES: apply a paste of sodium bicarbonate and water to the affected area.

- DIAPER RASH: sit the baby in a bicarbonate bath for a few minutes then pat dry.

- HEAT RASH: apply a paste of sodium bicarbonate and water.

- INSECT BITES AND STINGS: apply a paste of sodium bicarbonate and water. (Wasp stings respond to something acidic, such as vinegar; remember: B for Bicarbonate and Bee stings, V for "Vasp" stings and Vinegar.)

- SMALL BURNS OR SUNBURN: apply a paste of sodium bicarbonate and cold water to help prevent blistering and scarring. The cold paste takes heat from burnt skin; also, sodium bicarbonate dissolves endothermically (removing heat from its surroundings as it dissolves), which absorbs further heat.

- SPLINTER: apply a paste of sodium bicarbonate and water, cover with a sticking adhesive bandage and leave overnight to help draw out the splinter.

- SPLITS ON THE ENDS OF THE FINGERS: rub in a little sodium bicarbonate powder if the split is moist or make it into a paste first with a little water if it is dry.

# Stroke

Just as acidosis can encourage coronary heart disease and heart attacks (see pages 149–151), it can also encourage arterial disease and strokes ("brain attacks" caused by blood clots or bleeds from cerebral arteries).
An acid-producing diet is a possible contributory factor.

• Action:
Eat an alkali-producing diet (see pages 84–85).

An analysis of 8 studies involving 257,551 people found a lower risk of a stroke due to a clot or a bleed in those with a higher vegetable and fruit intake. The results suggest that consuming more than the recommended five helpings of fruit and vegetables per day is likely to cause a major reduction in the risk of a stroke.

*(The Lancet, 2006)*

## DID YOU KNOW?

The age-linked rise in chronic low-grade metabolic acidosis almost certainly encourages strokes. The levels of alkaline "buffer bicarbonates" in the blood begin falling at age 45, and on average are 18 percent lower by the age of 90. The blame lies with less efficient kidney function. So an alkali-producing diet becomes increasingly important with age.

# Tooth decay and gum disease

Mouth bacteria encourage tooth decay because they feed on food debris and produce acids that corrode protective tooth enamel.

Acidosis is another possible, though unproven, cause of tooth decay. This encourages the kidneys to excrete acidic anions. The alkaline minerals such as calcium that must accompany them have to be drawn from bones, organs, or even tooth enamel. Enamel thus weakened decays more easily.

Bicarbonate inhibits the formation of plaque (a sticky layer of food debris that can develop into tartar, which in turn encourages gum disease) on teeth. It also neutralizes acids produced by bacteria, helps prevent tooth decay, and increases calcium uptake by enamel.

- Action:
Clean teeth with bicarbonate-containing toothpaste or sodium bicarbonate powder.

# Tremor

This is a possible symptom of either respiratory or metabolic alkalosis. If you think your symptoms may result from rapid breathing (hyperventillation), try the action described in the box on page 163.

• **Action:**
Consult your doctor if it continues and you don't know the cause.

# Underactive thyroid

Chronic low-grade metabolic acidosis may depress the thyroid gland.

• **Action:**
Eat an alkali-producing diet (see pages 84–85).

Swiss researchers found that chronic metabolic acidosis reduced thyroid function by decreasing the thyroid hormones triiodothyronine and thyroxine, while increasing thyroid-stimulating hormones. They suspect that this might account for some of the effects of metabolic acidosis.

*(American Journal of Physiology, 1997)*

Two possible reasons for low thyroid-hormone levels are poor absorption of nutrients from the gut, and too little oxygen in the blood. Both are associated with chronic metabolic acidosis and perhaps, though this remains unproven, with the chronic low-grade metabolic acidosis commonly found in those who eat an unhealthy diet.

# A last word

It is sometimes suggested, though hasn't yet been proved
or disproved, that chronic low-grade metabolic acidosis
also encourages:

- Hyperactivity in children

- Lupus

- Multiple sclerosis

- Myasthenia gravis

- Pre-menstrual syndrome

- Sarcoidosis

- Schizophrenia

- Scleroderma

# Useful Websites and Suppliers

Ask a pharmacy or drug store to order sodium bicarbonate in amounts large enough for your domestic use.

Or order on-line: Remember that if you want baking soda for consumption either in food or as a remedy, you must ensure that the product you buy is "food grade."

### Arm & Hammer
www.armandhammer.com
Email via the website to find your nearest supplier, or buy the products at your nearest Walmart, Costco, Home Depot, or Lowe's. Supplies baking soda and various laundry, homecare, and personal- and pet-care products that contain this substance. For example, they produce a 20-oz FridgeFreezer Pack, which has non-spill vents on three sides of the box so it won't spill when you leave it open to freshen your refrigerator.

### Bob's Red Mill
www.bobsredmill.com
Email via the website to find your nearest supplier. Supplies baking soda and gluten-free baking powder in large amounts (4½-lb bags) for cooking, a boon and for anyone who is sensitive to the cereal protein gluten. This also a source of course "Scottish" oatmeal.

For anyone concerned about unproven suggestions that aluminum encourages Alzheimer's disease, baking soda produced in the US does not contain any traces of aluminum.

The following websites give helpful advice on healthy eating, weight management, and ways to improve nutrition.

www.choosemyplate.gov
The USDA website is packed with information and useful guidance on healthy eating and general nutrition with links to other useful websites.

**The Diet Plate**
www.dietplate.us

**The American Nutrition Association**
www.americannutritionassociation.org

www.preventcancer.aicr.org
Packed with information on healthy eating, weight management, and ways to reduce cancer and other health risks.

If you are keen to try making some of the beauty or household products, you may find the following useful:

www.saveoncitric.com
Suppliers of chemicals (including sodium sesquicarbonate), essential oils, and all you need to make soap and bath products, and much more.

# Index

and acid production 79, 80, 81, 82, 86, 92, 107–108, 129
and cancer 133
"folding" 95, 96
immune response to 108, 144–145, 146–147
*see also* high-protein diets
purines 79, 147
ptyalin (salivary amylase) 107

**R**
rapid breathing (hyperventilation) 105, 128, 131, 142, 163, 176
red blood cells 95
refrigerators, cleaning 16
restless legs 171
rheumatoid arthritis 129–130
rust removing 33

**S**
saliva 107
salt (sodium chloride) 8, 9, 119
in diet 79, 87, 158
saucepans, cleaning 18
scones 72
Scotch pancakes 74
scouring powder 14

secretin 109
self-rising flour 41
semolina cakes 58–59
shaving 117
singing hinnies 73
sink pipes, cleaning 16
skin pH 94
skin problems 172–173
skin-softening paste 117
slimming diets 140, 143
small intestine, acid-alkali balance 109–110
"soda loading" 160
soda water, homemade 123
sodium 95, 97, 99, 104, 107, 108, 110, 150
high blood 125
retention 104
sodium ascorbate 134
sodium bicarbonate 8–9
application to skin 126, 172, 173
dose 122–124
side effects 125
stomach production 107–108
taking internally 122–124
tips and warning 122–123
*see also* baking soda
sodium bicarbonate paste 126
sodium chloride *see* salt
sodium-sensitivity 153–154

splinters 173
sponges, cleaning 15
stain removal 24–25
clothing 26–28
stomach, acid-alkali balance 107–109
stomach (hydrochloric) acid 90, 108–109
low 109, 110, 153–154
stomach cancer 109, 154
stomach juice 94, 107
strokes 139, 174
strong acidic anions 78, 92, 96, 97, 109, 111
strong alkaline cations 78, 97 109, 111
strong-ion difference 79, 97
sugar 78
sugar shortage 142
sulfate 78, 92, 97, 99
sulfur 92
sunburn 173
sympathetic nervous system 97

**T**
tannin stains 24
tempura 53
thyroid, underactive 176
tissue fluid 91, 94, 97, 99
tooth decay 175

Notes

Notes